# The search for our beginning

*Frontispiece*   Discovery of the Allan Hills 76009 chondrite, Victoria Land, Antarctica. Note that the ice is almost completely free from snow cover. The photograph was taken by Dr E J Olsen, a member of the first expedition whose primary objective was the search for meteorites in an area where none had been discovered previously. The expedition was provided with helicopter transport by the US Navy.

# The search for our beginning

An enquiry, based on meteorite research, into the origin of our planet and of life

Robert Hutchison

British Museum (Natural History)

Oxford University Press

© Robert Hutchison 1983
First published in Great Britain by the
British Museum (Natural History), Cromwell Road, London SW7 5BD
and Oxford University Press, Watton Street, Oxford OX2 6DP, 1983

First published in the United States by Oxford University Press, Inc., 1983

**British Library Cataloguing in Publication Data**
Hutchison, R.
 The search for our beginning.
 1. Solar system—Origins
 I. Title
 521'.54    QB501    83-043007
 ISBN 0-19-858505-5 OUP (UK)
 ISBN 0-19-520435-5 OUP (USA)
 ISBN 0 565 00851 X BM(NH)

Printed in England by Butler & Tanner Ltd,
Frome and London

# Contents

# Acknowledgement

Research scientists are privileged to receive support from a public which generally is unable to gauge the results of its investment. This book was written to acknowledge such support. It is aimed at a wide readership in the hope that it will be enjoyably informative on recent advances in planetary science. As a meteoriticist I acknowledge also the efforts of mainly uneducated people from all over the world who assist in meteorite recovery.

I thank my colleagues Andrew Graham and Alex Bevan for reading and criticizing the first draft, and I thank Ed Olsen for encouragement at an early stage. I was delighted by the response of many friends and acquaintances to my request to use their material; they are acknowledged individually where appropriate. Finally, I thank my wife and daughter for their patience while I worked on the book, and my wife for helping me simplify the text, and for proof correcting.

# Introduction

We, each of us, have part of a star inside us. The calcium in our bones was made in a star and distributed by the nuclear explosion that ended its life before our Sun was lit. Evidence for such an event cannot be found on Earth because it is an active planet. Here, mountains are still being built, the Atlantic Ocean is ever widening, and the Earth's interior is hot enough to produce sporadic volcanism over much of the globe. Change has affected ocean basins and continental margins over most of the planet's surface and thus has erased their history before about 200 million years ago. Parts of the continents, however, have survived earth-processes much better than the oceans. But although we can trace the history of most continental interiors back to a major period of formation between about 2500 and 3300 million years ago, and, in a few cases, as bar back as 3800 million years, of earlier events we know nothing. On Earth the record stops at a time, 3800 million years ago, when liquid water was present and when rocks of continental type had already formed.

To look back beyond this time we can get a little information from the Moon, but essentially all of our knowledge comes from the study of meteorites, mostly pieces of rock but a few of metal, that fall to Earth from space. Many meteorites come from the break-up of small planetary bodies that were never as hot as the interior of the Earth; some of them have remained unaltered for about 4550 million years and so preserve evidence of the conditions under which they formed. A very few meteorites contain water and compounds of carbon, hydrogen, nitrogen and oxygen. These may be samples of the material that gave us our oceans and atmosphere and provided the stuff from which life evolved.

Meteoritics – the study of meteorites – is therefore relevant to the early history of the Earth and of its Moon and neighbouring planets. Meteorite research forms the common ground between astrophysics, geology, cosmochemistry, organic chemistry and astronomy. It is ironic that to look into the Earth's most distant past we must examine the most recent arrivals on its surface.

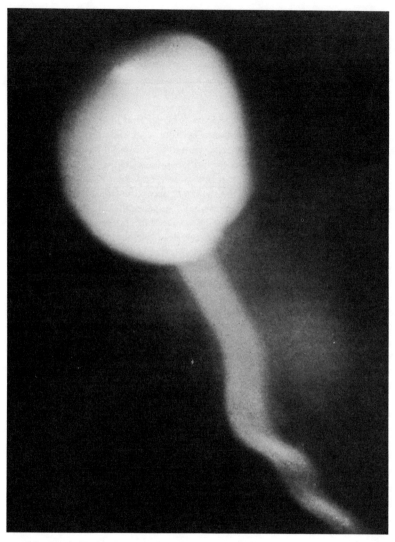

FIG 1.1    Fireball of 24 March, 1933, that produced the Pasamonte meteorite. Twisting of the dust trail indicates that the body changed direction. Courtesy Mr Glenn I Huss and the American Meteorite Laboratory.

# 1
# Debris from space

It is estimated that each year the Earth receives about 100 000 tonnes of material from space. This would obviously be catastrophic were it all to fall at once, but the great bulk arrives as meteors that burn up harmlessly in the upper atmosphere. Also, this present rate of bombardment is negligible when compared with the weight of the Earth, for 100 000 tonnes per year over 4550 million years represent only one ten millionth of the present weight. But five or six times every year a few witnesses are privileged to see the fall of a meteorite somewhere on Earth. The accompanying phenomena may be terrifying or awesome, so that primitive people usually either revere or destroy whatever is recovered. For example, the sacred stone of Kaaba, in Mecca, is reputedly meteoritic although it has never been subjected to scientific investigation, and several meteorites have been found in burials of North American Indians. At the other extreme, there are numerous reports of stony meteorites having been broken up by superstitious witnesses to their fall.

When a body enters the Earth's atmosphere from space, its minimum speed is 12 km per second (27 000 miles per hour), or a little more than the escape velocity. At such a hypersonic velocity, some forty times faster than sound, frictional heating by the atmosphere occurs on the front surface of the body even at heights of over 120 km (80 miles). At lower altitudes the still tenuous atmosphere is unable to move round the rapidly moving object, but builds up as a nosecap of compressed air. In order to survive to the Earth's surface, the object must be strong and not friable and it must be large enough for some material to be left at the end of the highly destructive flight.

Friction with the air causes an electric charge to build up in the air surrounding the rapidly moving object. This effect is analogous to the static produced when man-made fibres are rubbed. But the charged air produced by a potential meteorite can occupy a volume many metres across (FIG 1.1). A bright glow is emitted and a brilliant fireball results. Fireballs are often visible in broad daylight and may occasionally rival the Sun in brightness. The surface of the object is heated by the buffeting from the air, causing the outside surface to melt. The melted material is immediately swept away by the air, allowing no heat to penetrate into the interior, which remains cold. Behind the falling body the melted droplets solidify to produce a trail of fine dust that may persist for tens of minutes in the upper atmosphere. This type of melting and removal of material by the atmosphere is known as ablation. Irregularities in the shape of the body result in differences in the compressive force acting upon different parts of it, which usually causes the body to split. Except for very tough materials, pieces, large and small, are broken off and themselves are ablated and further fragmented. During the luminous flight, which may last from two or three seconds to almost one minute in the exceptional case of a very low-angled, almost grazing, trajectory, the body is being decelerated and consumed. The rate of deceleration increases as the body penetrates to the denser part of the atmosphere. By the time it is only 10 to 30 km up (6 to 20 miles), the potential meteorite will have been slowed down to a velocity at which its surface is no longer heated to fusion. Any melt still on the surface solidifies to a (usually) dark crust. The glowing envelope of electrically charged air shrinks and collapses and so the fireball is extinguished. The remaining material falls, darkly, under gravity to become a meteorite which, if observed by someone with connections with the scientific community, will be named after its place of fall.

Fireballs that penetrate deep into the atmosphere are normally accompanied by sonic booms which add to the startling effect. Break-up and irregular motions resulting from the irregular shape produce multiple shock-waves that sound like a series of sharp cracks followed by a thundery peal due to echoes. Witnesses often erroneously associate sonic effects with observed break-up of the fireball but, because sound travels at a fraction of the speed of light, a minute or more must elapse before any sound from a break-up can be heard. Because of their rarity it is seldom that meteorite falls are witnessed by trained observers and so descriptions are often endowed with preconceived ideas which prove to be false. However, there is one commonly reported feature that is still unexplained. A hissing noise sometimes accompanies or even precedes the first observation of the fireball. The noise cannot be related to the shock-wave but could somehow be related to the electric charge of the fireball itself.

Finally, as the fireball usually is extinguished at an altitude of some 10 to 30 km (6 to 20 miles) in the atmosphere, the last, dark part of the fall is rarely observed. And unless the falling object lands among people, or on a car or house, it can easily pass unnoticed. Because of the lack of scale, observers who see the fireball to its point of extinction often think that the object fell 'behind the haystack' or 'on the other side of the trees' and not the reality of 80 or 100 km away (50 to 60 miles). To pinpoint the probable area of fall even within a 20 km (13 mile) radius is often impossible and necessitates interviewing observers on either side of the flight-path and beyond the point of extinction of the fireball. This means that the investigator must be prepared to cover an area of 10 000 square kilometres (4000 square miles), or more, in attempting to narrow down the field.

But in 1959, by accident, a bright fireball was photographed by synchronized cameras designed for the study of meteors. Although the film was almost burned out by the brightness, the Czechoslovakian astronomers managed to obtain precise measurements of the speed and path of the fireball. The position and height of the point of extinction – the end-point – were calculated and, after making allowances for the effect of winds during the dark part of the flight, an area for searching for meteorites was delimited. This led to the finding and

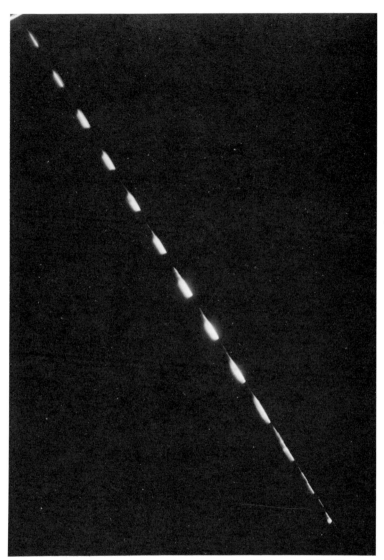

FIG 1.2    Fireball of the Innisfree, Alberta, meteorite, 5 February, 1977. The picture was taken by an automatic camera of the Canadian Camera Network which is designed for meteorite observation and recovery. A rotating wheel produces four segments per second. Fragmentation occurred towards the end of the path (lower right). Because the fragments travelled at different velocities, the light emitted was spread along the trajectory. Courtesy Dr I Halliday and the Herzberg Institute of Astrophysics of the National Research Council of Canada.

recovery of four pieces of stony meteorite out of a possible nineteen; the total weight was just less than 5 kg (11 pounds). The success of the photographic technique was noted by scientists in other countries. In addition to Czechoslovakia, networks of synchronized cameras were set up in the prairie states of the USA, in Canada, and over much of western Europe, including Britain. But after about ten years of operating, the prairie network of the USA closed down with only one successful meteorite recovery to its credit – Lost City, Oklahoma, which fell in January, 1970. The Canadian photographic network was successful with the recovery of the Innisfree, Alberta, fall of February, 1977, about eight years after operations began (FIG 1.2). So we have now photographic records of the atmospheric flight of three meteorites. In each case the data indicate that the bodies came to Earth on elliptical orbits from the zone between Mars and Jupiter occupied by the small, planetary objects known as asteroids (FIG 1.3).

Because meteorite falls are so spectacular, and each often has unique features that distinguish it from other falls, it is worth including a few eye-witness accounts. The first example here is that of the fall at Aldsworth, Gloucestershire, at about 4.30 in the afternoon of 4 August, 1835. Over several English counties bordering Wales a loud boom was heard, followed by a low rumbling sound that persisted for some seconds. At Cirencester in Gloucestershire a fireball was seen passing from west to east. It looked like a copper ball larger than an orange and had a tail or stream of light behind it; it made a rumbling noise like thunder which was heard by many people. And the inhabitants of the town marvelled that it should thunder on a serene day with a cloudless sky. At Aldsworth, about 20 km (13 miles) to the north-east of Cirencester, a meteoritic stone fell in a field and within 20 metres of workmen. They saw 'no unusual light, but heard the aerolite (stony meteorite) rush through the air, and felt it shake the ground by striking it with great violence'. It drove 'straw before it down into the earth for six inches [15 cm], till opposed by rock. When the men got it up it was not hot'. About 800 metres south of where the stone fell there was 'a shower of small pieces . . . Children thought it was a shower of black beetles, and held out their hands to catch them as they fell'. This account is from a summary published in 1852. The Aldsworth fall is unusual in that some of the spalled-off dust-trail landed soon after the meteorite. Another unusual feature is the absence of sonic or light effects close to the place of fall, which indicates that the flight-path was at a low angle to the horizontal and that the point of extinction of the fireball was at a low altitude. This is perhaps more clearly understood by considering the opposite case. A fireball with a steep flight-path will produce sonic booms directly beneath in the vicinity of the place of fall; likewise, eye-witnesses to the fall should also have witnessed the fireball, if only momentarily, in the sky above. But a low angled trajectory is more likely to have been obscured by intervening objects, and the shock waves which produce sonic effects may have been directed forward rather than towards the ground. The recovered stone weighed 600 grams and is now preserved in the British Museum (Natural History).

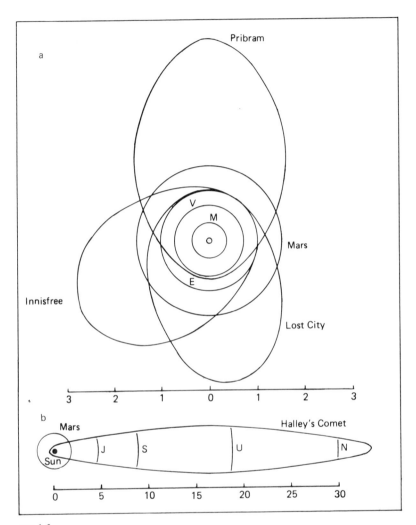

FIG 1.3

a) Orbits of the three meteorites which were determined photometrically. Note that all three are elliptical and extend outwards beyond Mars, but none intersects the orbit of Venus (V); Mercury's orbit (M) is also given. Note the scale is in astronomical units (AU), 1AU being mean distance of the Earth (E) from the Sun.

b) Orbit of Halley's Comet, relative to the outer planets Jupiter (J), Saturn (S), Uranus (U) and Neptune (N). Note the change in scale from Figure 1.3(a). Mars' orbit is included for comparison. Most comets have even more highly elliptical orbits; they may take thousands of years for a single orbit extending thousands of AU from the Sun.

A most spectacular meteorite fall was that of 24 March, 1933, which was visible over parts of the states of New Mexico, Colorado, Kansas, Oklahoma and Texas. The shower of meteoritic stones recovered over the following year is now known as the Pasamonte meteorite. The fireball appeared at 5 o'clock in the morning, moving along a low angled trajectory somewhat south of west. The light was so bright that it startled a ranch foreman who was indoors and about to sit down to breakfast. Luckily he had a loaded camera available, which he picked up as he moved outside to take the most impressive picture shown in Figure 1.1. These events were thoroughly investigated by Dr H. H. Nininger on whose detailed report this summary is based. Nininger personally interviewed witnesses over an area of about 300 000 square kilometers (300 × 400 miles) and hence predicted accurately the area where a search for meteorites might be fruitful.

Observers differed in their accounts, some having seen a single fireball, while others described a cluster or procession of fireballs; a few people witnessed fragmentation of a single fireball. It was stated also that the fireball changed direction. Although discounted by Nininger at first, this proved to be consistent with many observations, including the twisting of the dust-trail in Figure 1.1. Many people heard 'a swishing or whining noise at the instant of the fireball's passage', this being analogous to the unexplained 'hissing' mentioned on p. 10. Detonations and other sonic effects were heard two to three minutes after the passage of the fireball. A most unusual feature, of which several photographs exist, was a luminous cloud that persisted up to thirty minutes after the fireball had gone and before it could have been lit by the Sun. The cloud was some hundreds of kilometres in length and after the luminosity died down it continued to be observed for a further hour.

The descriptions so far relate to the rare occurrences in which some material survives atmospheric flight to land on Earth as a meteorite. Most objects, because of their friability, small size and weight or excessive speed, burn up in the atmosphere. These are called meteors or meteoroids. Some large, friable objects can produce bright fireballs but in general the term meteor refers to a more transient phenomenon seen as a streak of light in the night sky. Meteors are thought to be related to comets. This is especially so for the streams of meteors that appear to approach the Earth from particular parts of the sky at set times of year. Thus in the middle of August we have the Perseids, one of the most consistent annual meteor streams that gives rise to shooting-stars which seem to come from the constellation Perseus. Most meteors are produced by particles weighing much less than a gram and most are sporadic, meaning unrelated to meteor showers and appearing to come from random directions. However, photographic and radar observations indicate that meteors have highly elliptical orbits which have brought them from the outer reaches of the solar system, like comets (FIG 1.3).

Comets are thought to be either solid, 'dirty snowballs' or, less likely, swarms of dust and ice particles travelling together because of their mutual gravitational

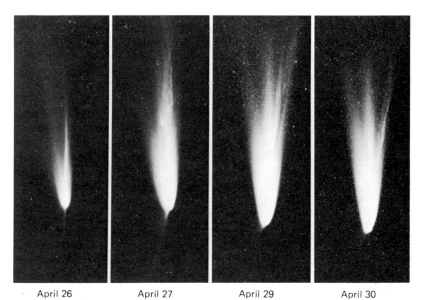

|  April 26 | April 27 | April 29 | April 30 |

FIG 1.4    Comet Arend-Roland of 1957. The comet appeared to emit a spike of bright material forwards and away from the tail on 26 April. Next day the spike was seen to be fan-shaped; turbulence in the tail is caused by interaction with the particles of the solar wind, which carry the Sun's magnetic field. The comet then returned to 'normal'. The spike consisted of dust grains driven by an explosive release of gas. Palomar Observatory photograph.

attraction. Comets come from the cold outer solar system and are composed largely of chemical compounds that are solid there but are gases on Earth. In addition to water ice, frozen ammonia, methane and carbon dioxide are probably predominant, and together form the 'snow'. When a comet passes close to the Sun it becomes heated and suffers intense bombardment by the atomic particles that are emitted by the Sun and are collectively known as the solar wind. Icy materials are turned to gas, sometimes explosively, causing a 'dirty snowball' to break up or a dust swarm to be disrupted (FIG 1.4). This phenomenon of break-up is therefore compatible with either theory of comets. After numerous orbits, each with a close approach to the Sun, collisions between dust-grains (the 'dirt' in the 'snowball'), the gravitational attraction of the planets, and other forces all tend to smear out the dust into a broad stream around the cometary orbit. Thus is a meteor stream born.

The Earth is therefore receiving dust from the outer solar system. Although most of this dust burns up as meteors in the atmosphere, it was predicted many years ago that the smallest particles reach the Earth's surface intact. (Technically speaking, these are micrometeorites.) The smaller an object, the larger is its

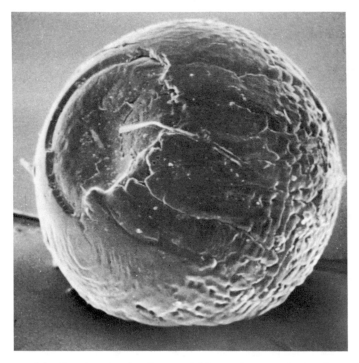

FIG 1.5   Cosmic spherule, diameter 0·07 mm, from a red clay almost 6000 metres beneath the surface of the Atlantic. The interior is iron-nickel metal with 61 per cent nickel, covered with a thin skin of oxides of iron. Courtesy Dr D W Parkin, University of Bath.

surface area in proportion to its volume. A particle about one twentieth of a millimetre, or less, in diameter has a surface area large enough to radiate away frictional heat during passage through the atmosphere. Such a particle will decelerate and float gently down to Earth, completely unscathed. In the past five years, micrometeorites have been identified in clay samples from the deep ocean (FIG 1.5); others have been collected in the atmosphere by high-flying aircraft. Laboratory study indicates that these may not all be solidified melt products from the ablation of meteorites. It is therefore probable that man has access to small samples of stony dust, with a little iron-nickel metal, that come from part of the solar system beyond the outermost planets.

FIG 1.6   Impact pit of the Jilin, China, chondrite (a) following the discovery and (b) after excavation. The pit was formed by a single stone of weight 1·77 tonnes belonging to a shower that fell on 8 March, 1976. This stone is the largest known. Courtesy Dr Ouyang Ziyuan and the Institute of Geochemistry, Academia Sinica.

We have now dealt with the fall to Earth of objects weighing between a fraction of a gram and about ten tonnes, the maximum weight of most meteorites. When a tough, mechanically strong body weighing considerably more than ten tonnes arrives from space it is only partially slowed down by the atmosphere. If the final weight is more than one tonne, the object will normally bury itself in a pit some two to six metres in depth (7 to 20 feet, FIG 1.6). This type of impact pit is merely a punched hole. Even heavier bodies produce impact craters, with earth and rock thrown out in all directions. An example of this is the fall of 12 February, 1947, in the Sikhote-Alin mountains of eastern Siberia. In this case the body broke up in the atmosphere and the various pieces of iron-nickel metal produced craters up to 26 metres in diameter. Although the pieces of meteorite may be broken up in an impact of this size, the impacting object is never destroyed. However, on the exceedingly rare occasions when an even larger and heavier body lands, the destruction to the Earth's surface is considerable. It must be stressed that no very large impact has occurred in historical times – meteorite impact poses an unlikely hazard to civilization.

When a body weighing more than about a hundred tonnes hits the surface of the Earth it produces an explosion crater. Friction during atmospheric flight results in only a slight reduction in the extra-terrestrial velocity, and impact takes place at hypersonic speeds of five or ten kilometres per second. The body penetrates the Earth's surface like a bullet, drilling a relatively small hole, but in so doing it is slowed down. During the very short time of deceleration, the energy which the body possesses by virtue of its great velocity (the kinetic energy) is converted to heat. The impacting meteorite is vaporized beneath the Earth's surface. In the cases studied so far, this means that most of the meteorite is converted to a bubble of iron-nickel gas under pressure. The bubble bursts, blowing its rocky cover upwards and outwards and causing it to fragment and turn upside down (FIG 1.7). The bedrock is impregnated with a trace of the meteorite which may be sufficient to allow scientists to identify the type of meteorite that produced an impact crater millions of years ago. In the formation of craters of this type, the energy released is considerably greater than in any man-made nuclear explosion up to the present. Obviously, the meteoritic projectile is almost completely destroyed in the process of forming an explosion crater. In the collision with the surface of the Earth, a shock-wave passes from the front to the back of the impacting body. This causes compression at the front, but when the shock-wave reaches the back it produces expansion and pieces are flung far and wide from the back surface. The main body of the projectile penetrates the land-surface and is vaporized. It is therefore normal for no meteoritic material to be recovered within an explosion crater; surviving material is found on the crater rim or in the surrounding area.

Our best and most easily accessible example of an explosion crater is Meteor Crater in northern Arizona (FIG 1.8). It was discovered by white settlers in the 1870s and very soon became a subject for controversy: were the lumps of meteoritic iron found around the crater associated with its formation, or was

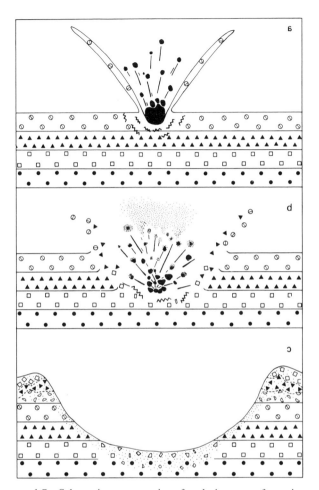

FIG 1.7   Schematic representation of explosion crater formation.

   a) Projectile strikes land-surface. Shock-wave penetrates downwards, jets of displaced material are formed. In the projectile, a shock-wave passes from front to back, scattering fragments from the back.

   b) The projectile buries itself and is decelerated. It heats and vaporizes, excavating the crater and pulverizing the underlying rock.

   c) A raised rim is built from some of the excavated material. In horizontal strata (e.g. Meteor Crater), the sequence in the rim is inverted and the strata are disrupted. The country rocks are impregnated with vapour from the projectile (stippling). Beneath the crater the rocks are fragmented, and beneath large craters they may be partially melted. Diameter of the crater is perhaps one hundred times that of the projectile.

FIG 1.8    Meteor Crater (Barringer Crater), Arizona, from the air. The crater is 1·2 kilometres across and 170 metres in depth, from the top of the raised rim to the floor.

their presence merely coincidental? Originally the latter was thought to be the case, for within some tens of kilometres of the crater are volcanic fields with many craters, all of which are undoubtedly volcanic in origin. G. K. Gilbert of the US Geological Survey was persuaded that some kind of steam explosion had created the crater, which is 170 metres from rim to floor and over a kilometre in diameter. In spite of Gilbert's recognition of the craters on the Moon as impact features, he could not bring himself to accept such an origin for the Arizona crater; the meteoritic iron was assumed to be unrelated to the structure. Such was the position towards the end of the nineteenth century.

As time went on more people became convinced that the crater had resulted from the fall of an abnormally large iron meteorite. An engineer called D. M. Barringer believed that the main mass of the projectile still lay buried beneath the crater and that, once located, it could be exploited as a source of iron and nickel. A company was formed and attempts were made to locate the object by drilling – no easy task in the desert of northern Arizona. The main difficulty encountered was not, however, lack of water, but an abundance of water at a level about sixty metres below the floor of the crater. Numerous holes were drilled and several shafts dug; the last shaft was sunk in 1927 from outside the crater rim but it too, had to be abandoned because of flooding. A scientist named Moulton was commissioned to estimate the size and weight of the body which had made the huge hole in the ground. He calculated that the weight might have been as low as 50 000 tonnes, or less than a hundredth of the estimate on which hopes of exploitation had been pinned, and that the iron and nickel must have been dispersed in a crater-forming explosion. So, by 1929, the correct inference was made that the crater was valueless as a source of metal. Financial support was withdrawn and apart from a few geophysical surveys no further work was undertaken. A rusting boiler and winding gear are all that now bear

FIG 1.9   Meteor Crater floor, November 1972. Industrial archaeology. Mr C. F. Lewis, of Arizona State University (right) and Dr F. Burragato, University of Rome.

witness to the feverish activity on the crater floor some fifty years ago (FIG 1.9). Subsequently, in the age of the motor car, Meteor Crater appears to have achieved commercial success as a tourist attraction.

Following the investigation of Meteor Crater several other features were recognized as a the result of explosive meteorite impact. The Wabar craters in Saudi Arabia and the Henbury craters in the Northern Territory of Australia were scientifically the most important. In each case pieces of iron meteorite were found in the vicinity as direct evidence of their impact origin. But in addition was noted the occurrence of slaggy-looking glassy rocks that were particularly abundant around the Wabar craters. L. J. Spencer and M. H. Hey of the British Museum (Natural History) showed that the glassy rocks had been produced by the intense heat of the meteoritic impact. This had caused the desert sand to melt and froth; chemical analyses of the meteorite and glassy rocks indicated that the latter contain iron and nickel in the same proportion as in the meteorite, whereas unaffected sand was found to be devoid of nickel. This was a practical demonstration that the impacting body had vaporized and that the vapour had impregnated desert sands fused in the explosion.

After the Second World War, aerial photography and the increased pace of geological exploration resulted in the discovery of many more impact features over most of the Earth's land surface. And with the advent of space exploration it was realized that planet Earth must once have been as heavily bombarded as the Moon and Mercury. But because the Earth is still active, most of the early

FIG 1.10   Shatter-cones from the Kentland, Indiana, structure. Apices of cones point towards the source of the shock that produced them. In the figure, the apices of two cones point upwards. The structure consists of deformed and overturned Ordovician and Silurian strata (500–400 million years old) which are poorly exposed beneath a cover of Carboniferous strata (300 million years old). There is no absolute proof that an impact caused the deformation which clearly occurred before the Carboniferous strata were deposited. Width 7 cm.

impact scars have been erased. One ancient impact produced a nickeliferous body of exceptional value: this is the sulphide ore deposit at Sudbury, Ontario. Although an extra-terrestrial origin for the nickel is still hotly disputed, it is generally accepted that the deposit came into being after a large impact about 1700 million years ago. The Sudbury structure is 59 × 29 kilometres in area and hence the weight of the impacting body must have been very much larger than that which produced Meteor Crater.

Nowadays impact features large and small are recognized over most large tracts of ancient rocks such as the Canadian Shield or the Siberian platform. But to establish that a circular feature on an aerial photograph is an impact crater necessitates a ground survey. In the absence of fragments of meteorite, evidence of intense shock must be sought. For example, the mineral, coesite, will form only at a pressure of some tens of thousands of atmospheres by the compression of quartz, which is chemically identical. Coesite has been recovered from Meteor Crater. Another feature diagnostic of intense shock is that of shatter-cone, the shearing of rocks along cone-shaped fractures, with the noses of cones always pointing towards the point of impact (FIG 1.10). This structure is found in various fossil meteorite craters but is absent at Meteor Crater, probably because erosion has not yet penetrated far enough down to reveal the shatter-cones which are expected to be there. In the absence of pressure indicators, verification of a circular feature as an impact crater may require drilling. Beneath the floor of a true impact crater is a layer of pulverized and partially melted rock formed by the intense pressures and temperatures that obtained directly under the decelerating projectile. Such layers have been proved in several Canadian examples and in the Nördlinger Ries structure of southern Germany.

Doubt is still cast, however, on the likelihood of an impact origin for some explosion structures because they lie along lineaments defined by the presence of features of undoubted igneous origin. There are today a small number of scientists who argue, with justification, that we do not yet understand all of the processes of volcanic activity and that the number of recognized 'impact' features has been over estimated.

So far we have discussed the explosive impact of solid, massive objects on the Earth's surface. One major natural explosion appears to have been caused differently. This was the huge detonation in June, 1908 in the region of the Tunguska River of central Siberia; this event has given rise to varied and sometimes erroneous speculation over the past half century. The explosion took place at 7.17 a.m. local time and although eye-witnesses' reports were submitted to the meteorological observatory at Irkutsk, no investigation was begun until 1921. Witnesses saw a 'blindingly bright' fireball, brighter than the Sun, crossing the sky from south-east to north-west. Afterwards there remained a very thick dust-trail which looked like 'a gigantic column'. Near-by observers saw fire and a cloud of smoke over the place of fall. There followed deafening detonations audible from over 1000 kilometres away. Ground tremors were felt over a rather smaller area; these had the effect of a minor earthquake, breaking windows and shaking buildings. At a trading post 60 kilometres from the fall one person who had been sitting at the time was thrown several metres and knocked unconscious. Before passing out he had 'felt heat radiating from the place of fall'. Forty kilometres south-east of the explosion a tent was lifted into the air, together with its occupants. The compression wave in the atmosphere travelled twice round the Earth and was recorded widely over north-western Europe, but the records were not discovered until many years after the event, as were similar records of ground tremors.

A search for the place of fall was initiated in 1921 and was continued intermittently over the next five years. Success was achieved only in 1927 after reports of devastated forest had been received from nomads during the previous year. The 1927 expedition was led by L. A. Kulik, who was amazed to find uprooted forest over an area up to 30 kilometres across. Here the trees lay with their roots towards the centre and their barren tops pointing away from the point of impact. An exception was in the central part, where the trees still stood albeit with broken tops and bare of branches. Much of the southern portion of the area was found to be swamp and at first it was thought that the swamp had resulted from the impact. Careful study, especially during post-war expeditions, disproved this theory. All attempts to locate meteorites in the area were fruitless, but in 1957 'meteoritic dust' was discovered in soil samples collected during the pre-war expeditions. And there the matter rests.

The most generally accepted explanation of the Tunguska phenomenon is that it was caused by a small comet that collided with the Earth after having passed behind the Sun. Such an object would have remained invisible to astronomers during its approach, the last stages of which were out of the morning Sun. However, the tail of a comet always points away from the Sun, so if the cometary theory is correct, the tail should have encountered Earth first. Contemporary records show that over western Russia, from the Caucasus to Siberia, and over the rest of Europe the night after the fall (30 June) was abnormally bright. Indeed, without artificial light, newspapers could be read at midnight. This is consistent with the scattering of light in the upper atmosphere by fine dust from a comet's tail, which presumably entered the atmosphere over Europe before the impact of the main body to the east. Subsequently, dispersal of the debris throughout the atmosphere caused the nights to be bright until the end of August, 1908. Finally, laboratory simulation by Russian scientists indicates that the body exploded in the atmosphere almost ten kilometres up; that it did not actually strike the Earth's surface. The Tunguska event can therefore be explained in normal scientific terms without invoking anti-matter, black holes or alien space-travellers. But its uniqueness in man's experience should be a lesson that we have not yet witnessed every phenomenon in nature.

Although cratering on the scale of Meteor Crater occurs, on average, somewhere on the Earth's surface once every 1300 years, very large impacts are statistically much less frequent. In spite of their rarity such catastrophic events may have been important in producing climatic change, which, albeit temporary, could have determined the path of evolving life. It is estimated that approximately once in a hundred million years an asteroid of 5 to 10 kilometres diameter lands on Earth. The most recent encounter probably occurred 65 million years ago. Sediments of this age, at the boundary between the Cretaceous (older) and Tertiary geological periods, are enriched in platinum and other noble metals. The enrichment is world-wide, occurring even in continental sedimentary sequences. The argument is that only disintegration and vaporization of an asteroid could have caused the distribution of platinum-group metals over the contemporary surface of the Earth.

The Earth has a dense core (see FIG 2.5) which most likely formed over 4000 million years ago when the planet was largely molten. Dense iron-nickel metallic liquid drained to the centre to form the core, taking with it the chemical elements which dissolve preferentially in molten metal. Among these elements are gold, platinum and iridium. Not surprisingly the rocks of the Earth's crust are highly depleted in noble metals compared with either iron meteorites or stony meteorites of types which have not been melted. Because the terrestrial levels are so low it appears that the major enrichment in noble metals in the sediments at the Cretaceous-Tertiary boundary must have had an extra-terrestrial source – an asteroid or perhaps a comet. Iridium is the noble metal most easily determined and in the literature this is often the only one mentioned.

Impact of a 10-kilometre asteroid or comet would produce a crater many hundreds of kilometres in diameter, which, if extant today, ought to be readily identifiable. That no fresh crater has been found suggests that impact occurred in oceanic crust which has since been destroyed by subduction (see FIG 3.1). One possible location of the impact was in the north Pacific where most of the crust formed 65 million years ago has already been subducted. The impact would have blown a hole in the atmosphere through which vaporized ocean would have been drawn, and the enhanced abundance of water vapour in the atmosphere would have trapped heat from the Sun, and so caused warming. The climatic change could have been responsible for killing off the dinosaurs and for other roughly contemporaneous extinctions among marine organisms. With the demise of the dinosaurs the way was open for the evolutionary diversification among the mammals, which led to the birth of ourselves. Asteroidal impact may be a mixed blessing, but its potential effects are of undoubted significance.

Because meteorite falls are mainly random, the five or six observed falls per year represent only a small portion of the total number of meteorites that the Earth receives annually. About three-quarters of all meteorites must land in water and of the remainder most will pass unobserved as a result of their falling in hot or cold desert areas or in dense forest. Even in populated areas many falls are probably unnoticed when cloud cover obscures the fireball. In recent years there have been several reports of meteorites, the presence of which was discovered only through damage to buildings. Estimates of the total number of falls have varied, but the recently obtained figure of about 3500 annually seems to be well supported.

The fate of an unobserved meteorite depends largely on location or climate. No meteorite has yet been recovered from the deep ocean or from water more than a few metres deep, and only one fossil meteorite has been found in deposits more than a few tens of thousands of years in age. This is largely because the most abundant minerals in meteorites react with a moist, oxidizing atmosphere; the simplest example is the conversion of iron-nickel metal to rust. Under most circumstances, then, meteorites disintegrate in a time which is short compared to the rate of infall. Notable exceptions are hot and cold deserts, such as occur in the Arabian peninsula and Western Australia, and in Antarctica. There have been few observed meteorite falls in the hot deserts and none in Antarctica, but Western Australia in particular has yielded an abundance of meteorite finds over the past fifty years or so. Antarctica is very much a special case. Here, meteorites which land on the ice-cap may be kept in cold storage for long periods during the build-up of ice. Movement of the ice can subsequently transport the meteorites to areas where the ice is forced upwards to be ablated (blown away) by the fierce winds. This leaves behind the cargo of meteorites which become concentrated on the erosion surface of the ice. In the Yamato Mountains of north-east Antarctica, Japanese expeditions since 1969 have recovered over four thousand pieces of meteorite. Following the example of the Japanese, in 1976 scientists from the USA began to search for meteorites on the opposite side of eastern Antarctica. They too met with success (see Frontispiece). The total number of fragments of meteorites recovered from the two areas is now about 5000. Because the average size of the fragments is much smaller than for observed falls, it is estimated that the number of separate meteorite falls represented is only about one-two-hundredth of the number of fragments. (The small size of the fragments is thought to result from fracture either by freeze-thaw or by compression in the ice.) Thus, over about ten years, two areas of Antarctica have, between them, yielded specimens of about 50 meteorite falls. This constitutes over 2 per cent of the 2100 meteorites from other locations which found their way into scientific collections over the preceding 200 years or so. Of the 2100 meteorites, some 40 per cent are observed falls.

We have discussed meteors, comets and meteorites in terms of modern concepts in planetary science. However, less than 200 years ago scientists did not accept that stones could fall from the sky. But, once freed from this misconception, science grew in leaps and bounds and the study of meteorites was linked with advances in chemistry and physics. These form the basis of the next chapter.

# 2
# The historical perspective

Before iron technology had come into being, man gratefully, if sometimes unwittingly, used iron from beyond the planet Mars. Meteoritic iron must then have been one of the most prized commodities in the world. Artifacts of meteoritic iron are known from tombs in North America, China, the Middle East, and Egypt; and in the last at least, it is clear that the ancients knew that iron came from the sky. The hieroglyph (FIG 2.1) which is now used for the cover of the journal *Meteoritics* is translated as 'heavenly iron'.

The great value placed on meteoritic iron is well illustrated by the occurrences at Cape York, north-west Greenland. 'Cape York' is the name given to a shower of mostly very large lumps of iron meteorite that were worked by Eskimos in prehistoric times. In this region, 'prehistoric' means before the arrival of European explorers in the early part of the nineteenth century. When Captain John Ross visited the area in 1818 he found the Eskimos using bone harpoons and knives tipped with flattened pieces of iron (FIG 2.2). Almost eighty years later, the American explorer Peary recovered three masses of iron, the largest weighing 30 tonnes. These he sold to a purchaser for presentation to the American Museum of Natural History, the proceeds going to finance later expeditions.

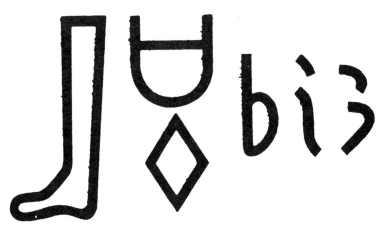

FIG 2.1   Egyptian hieroglyph, probably pronounced 'bith', signifying iron, especially iron from the sky. Courtesy of *Meteoritics*.

FIG 2.2   Eskimo knife presented to Captain John Ross during his visit to north-west Greenland in 1818. The cutting edge is a series of flakes of iron-nickel from one of the irons of the Cape York shower, set in bone. Width 20 cm.

Around the site of each mass of iron were mounds of stones used by the local people to try to pound fragments from the tough parent irons. Because the nearby rocks were friable, boulders of dark basalt were carried over 30 kilometres by the Eskimos for use as hammers. Some mounds of shattered and battered stones still testify to the countless man-hours which the Eskimos worked to secure an iron tip for a harpoon. The 'pilgrimage' of the people of north-west Greenland must have been an annual event for many centuries. Such was the price paid for meteoritic iron by primitive peoples.

There are sparse records of meteorite falls from ancient Egyptian times until the Middle Ages, but the earliest fall of which material survives did not take place until 1492. On 16 November of that year a large stony meteorite was seen to fall near Ensisheim, in Alsace (FIG 2.3). The bulk of the stone is still preserved in the town hall there and fragments are distributed around the world's museums. Thirty-three years later, the meteorite fall that caused most destruction to man is said to have occurred. One stone reportedly fell on the Castle of Milan, setting fire to the munitions. However, this story seems unlikely. During these times and, indeed, until the close of the eighteenth century, people did not believe that meteorites came from outside the Earth. It was thought that stones were carried up into the clouds by water spouts and then redeposited during thunderstorms. In spite of observations of meteorite falls from clear skies, the association of 'stones from the sky' with thunder persisted into the nineteenth century. For example, Congressional certification of the Weston, Connecticut, fall of 1807 was requested by several eye-witnesses; and President Jefferson, in true politician's language stated '. . . It may be difficult to explain how the stone you possess came into the position in which it was found. But is it easier to explain how it got into the clouds from whence it is supposed to have fallen? . . .' But with the birth of physics and chemistry, scientific times were changing.

Von dem bonnerstein gefallē jm ʳ̃c̈ij.iar:vor Enſiſhein.

FIG 2.3   Wood-cut depicting the fall of the Ensisheim, Alsace, meteorite in November 1492. Note the (erroneous) suggestion of thunder-clouds and lightning. This is the oldest observed European fall of which material survives.

In 1794 a German physicist called Chladni published a book on the origin of certain masses of iron. Although the occurrences of iron were from widely spaced localities, he argued that they were so similar in appearance that they must have a common origin. Because they were unrelated to their surroundings, Chladni argued that the irons were extra-terrestrial in origin. He then went on to speculate that the irons had come from a disrupted planet. On flimsy evidence Chladni had reached the correct conclusion some six or eight years before its proof and general acceptance by scientists.

Towards the end of the eighteenth century, rock and mineral collecting was a fashionable pastime for scholars and the rich. Meteorites were sometimes included in these collections and several irons, stony-irons and stones were quickly distributed throughout Europe (FIG 2.4). Such materials soon became objects for study by the early chemists. In the 1770s, for example, a group of scientists of the French Academy, including Lavoisier, performed a chemical analysis of the Lucé stony meteorite. They found it to have a considerable sulphur content and concluded that it had been formed by the action of lightning on ordinary, terrestrial, iron pyrites. But, intrigued by various eye-witness accounts of stones being associated with 'fiery meteors', an English chemist named Howard separated the metallic iron component which is present in almost all stony meteorites. He found that it contained nickel, as did iron meteorites. Here was the link between irons and stones, the proof of similarity which Chladni had lacked.

Finally, in 1803, at L'Aigle, France, there occurred the fall of a meteorite shower that was witnessed by a member of the French Academy. The fall had undisputed authenticity; stones occasionally *did* fall from the sky.

The next major step forward for planetary science came in 1830, when

FIG 2.4   One of the stones of the meteorite shower that fell at Siena, Italy, in 1794. It was thought originally to have been erupted by Vesuvius, almost 400 kilometres distant. Width 6 cm.

Charles Lyell put forward the principle of uniformitarianism in his book, *Principles of Geology*. Lyell's idea was that, geologically speaking, the present is the key to the past; that the geological processes at work today are the same as those that deposited the strata and folded the rocks during past eras. Lyell's principle demanded an Earth of infinitely great age. This concept was later supported by evolutionists such as Charles Darwin, who required lengthy periods of time during which natural selection could take effect. Thomson, who became Lord Kelvin, was the leading antagonist to the idea of an Earth of limitless age. He argued that if our planet had originally been molten, it would have cooled in twenty million years – unless it had a then undiscovered source of heat. For almost the whole of the second half of the nineteenth century the controversy raged, T. H. Huxley being Kelvin's leading opponent. But a few years before his death in 1907, Lord Kelvin reputedly admitted defeat when faced with the evidence of the newly discovered radioactivity, the source of heat which slows down the cooling of our planet. However, he never recanted in print and his last articles were fiercely defensive.

The discovery of radioactivity opened up new avenues of scientific exploration for the geologist as well as for the physicist. By the early 1930s a very few perceptive geologists or physicists of broad outlook were ready to make a serious attempt at determining absolutely the ages of rocks from ancient parts of the Earth's crust. It was known then that uranium decays at a fixed rate to the element lead. To determine the ages of ancient rocks, all that had to be done was to locate uranium minerals in them and then measure the uranium and lead contents of the minerals. The ratio of these two chemical elements is proportional to the time since the formation of the mineral – provided that the mineral contained no lead when it was formed and provided that no uranium or lead had been added or lost since the time of formation. These are fairly reasonable assumptions for a uranium mineral such as uraninite (chemical formula $UO_2$-$U_3O_8$), and estimates of the age of formation ranged up to 1460 million years. This was, of course, a minimum age, for the rocks in which the mineralization took place must have been older still. And although this estimate is only 30 per cent of the modern estimate of the earth's age, 1460 million years is certainly closer to the truth than Lord Kelvin's 20 million years. Geologists had been given a time-scale long enough to satisfy their estimates of the rate at which strata are deposited and the rate at which evolutionary change is likely to occur. It is to the credit of the geologists that, as long ago as 1893, it was estimated from the total thickness of fossiliferous strata that the first animals with shelly outside skeletons lived about 600 million years ago. With improved techniques over the past century, this figure has decreased by only 5 per cent to 570 million years.

At about the same time as the first attempts to utilize radioactivity to measure the Earth's age, the science of geochemistry was being born. For although the chemists of the eighteenth and nineteenth centuries must be commended for their attempts at analysing rocks, minerals and meteorites, an understanding of the chemistry of the Earth as a planet was not achieved until the 1920s. V. M. Goldschmidt, by examining the distribution of chemical elements between different minerals, especially in meteorites, established that, for example, some metals such as gold almost always occur as a metal on their own or in certain other metals. He found that in rocks from the Earth's crust, gold is much rarer than in iron meteorites, and went on to propose that most of the Earth's gold is in an iron-nickel core (and hence inaccessible to man). That the Earth has a dense, probably metallic, core had previously been proposed by physicists studying earthquake waves (FIG 2.5), and also because of the analogy with iron-nickel meteorites which were thought to represent the cores of asteroids. Goldschmidt classified the chemical elements according to their chemical affinities. On Earth, lead often occurs as the sulphide, galena. Elements such as lead, with an affinity for sulphur, he called 'chalcophile'; elements such as gold, with an affinity for metal, he called 'siderophile'; elements such as silicon and magnesium which form oxides (or stony minerals), he called 'lithophile'; and finally, elements such as nitrogen or neon, which remain as gases, he called 'atmophile'. This classification implicitly leads to a theoretical model for the

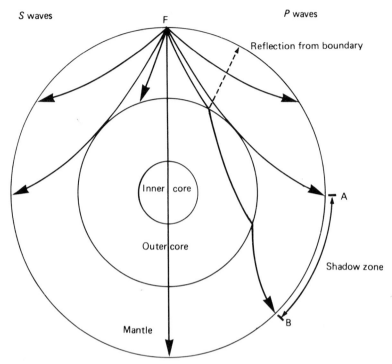

*S* waves
*P* waves
F
Reflection from boundary
Inner | core
Outer | core
A
Shadow zone
B
Mantle

FIG 2.5   Internal structure of the Earth deduced from the passage of earthquake waves. An earthquake at F (focus) produces two sets of waves that propagate in all directions. Waves that vibrate at right angles to the direction of motion (*S*, or shear, waves) cannot pass through liquid. Because they do not pass through the Earth below a certain depth, it is inferred that a liquid core exists. In contrast, the waves that vibrate along the direction of travel (*P*, or pressure, waves) do pass through liquids. These waves are analogous to ripples that spread outwards when an object is thrown into water. A *P* wave that grazes the core reaches the surface again at A, but one that just enters the core is refracted downwards, because the outer core is liquid, although denser than the mantle above, and does not reach the surface until point B. Between A and B no earthquake waves are received from a quake at F; this area is known as the 'shadow zone'. A *P* wave passing vertically down from F goes straight through mantle and core. From the properties of *P* waves it is inferred that a solid inner core exists.

distribution of the chemical elements on Earth, and thereby established the branch of science called geochemistry – the chemistry of planet Earth. The name is now more widely applied to include the Moon, meteorites and planets. Because of differences in bulk chemistry between the Earth and meteorites or other planets, the geochemical classification of an element may change from one location to another. For example, the atmosphere of Venus contains sulphuric acid, in which case sulphur is partly atmophile as well as chalcophile.

Following the Second World War, advances in technology and theory led to advances in planetary science. One of the most important which has stood the test of time is the estimate of 4550 million years for the age of formation of many meteorites and of the Earth. In the early 1950s, C. C. Patterson of the University of California extracted lead from various meteorites. Now the chemical element lead has atoms of four different types; although their chemical properties are the same, the atoms have different weights and are called isotopes of the element, lead. Two of the four isotopes are produced by the radioactive decay of uranium at different rates. What Patterson did was to measure the ratio of the two isotopes of lead formed by uranium decay. This ratio is proportional to the time elapsed since lead began to accumulate from the decay of uranium; the length of time is essentially the age of the meteorite. Once again, this technique depends on the assumption that no lead has leaked from the meteorite. Patterson then established that lead from modern sediment in the Pacific Ocean has much the same ratio as in meteorites, implying that the Earth has the same 4550 million-year age. More recently, the rate of uranium decay has been more precisely determined, but the best estimate of the Earth's age is still close to 4550 million years. It must be emphasized that we know the age of the Earth only indirectly and that our estimate is tied to the idea that meteorites and the Earth have a common origin and were formed from materials with a similar history.

As man entered the space age, the next important development in the understanding of our planet was the virtual proof that some of our oceans are widening; that Europe and Africa are moving away from the Americas. Although the idea of 'continental drift' had been put forward as early as 1910 by a meteorologist called Wegener, it was held in abeyance for fifty years, mainly because of the opposition of physicists. The problem was that physics was unable to provide a mechanism capable of producing the vast forces required to move continents. But in the 1960s physics did eventually supply the proof. Survey work had earlier shown that a system of ridges runs throughout our oceans. The new measurements showed that on either side of each ridge the sea-floor consists of strips magnetized in opposite directions (FIG 2.6). There is an alternation of strips magnetized in the present direction of the Earth's magnetic field with strips showing the exact opposite magnetization. In the Atlantic, the pattern of strips is symmetrical about the feature, known as the mid-Atlantic ridge, which is some 12 000 kilometres (8000 miles) long. The explanation is that the Earth's magnetic field alternates from time to time, which means that the north magnetic pole changes places with the south magnetic pole. The mechanism is still under debate. The strips are produced by the upwelling of molten rock at the middle of the ridge – the volcanoes of Iceland are an example of this. As the molten rock cools, it becomes magnetized according to the prevailing direction of the Earth's magnetic field. New molten rock coming up from depth at the middle of the ridge pushes the earlier formed rocks to each side, thus breaking them into two strips, one on either side of the

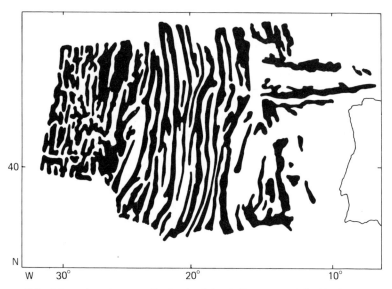

FIG 2.6   Magnetic strip anomalies in the Atlantic floor west of the Iberian Peninsula. Black: normally magnetized rocks, that is, rocks that solidified from melts when the Earth's magnetic field was in its present configuration. Blank: reversely magnetized rocks that solidified when the Earth's field was reversed. The overall pattern is of alternating north–south strips pushed eastwards from the mid–Atlantic ridge (see text). To the west of the ridge a similar pattern exists, indicating that the crustal structure of the Atlantic is roughly symmetrical about the mid–ocean ridge. In the Bay of Biscay, E–W trending anomalies mark the site of an extinct spreading axis. Courtesy Dr E. J. W. Jones after figure by C. A. Williams.

ridge. It is this upwelling, with production of new rock in the middle of the Atlantic, which is apparently pushing the continents apart. For about 160 million years the process has gone on, the resulting ocean having grown at the average rate of a few centimetres per year. Although the driving process is not yet understood, geologists and physicists have suggested several possibilities. They had to. For the evidence is so compelling that the reality of 'sea-floor spreading', as the phenomenon is now known, is virtually undisputed by scientists.

Then came 1969. This was *the* most outstanding year in the history of planetary science. During a single orbit of the Earth round her Sun occurred four events each of which had a great impact on the direction of scientific research. And all save one of the events was completely unforeseen.

On 8 February there fell in Mexico a large shower of meteoritic stones of a very rare type. Over two tonnes of material quickly found their way into scientific collections, allowing researchers access to large quantities of a kind of

FIG 2.7   One stone of the Allende, Mexico, meteorite shower. FC: fusion crust. CH: chondrules protruding from a fractured, interior surface. CAI: calcium–aluminium–rich inclusion (see Chapter 7). Width 11 cm.

meteorite which had previously been available only in gram-sized amounts. In the Allende, Mexico, meteorite are white inclusions in which was discovered evidence for the formation of chemical elements in the runaway nuclear reaction which is thought to have taken place in an exploding star. Research is still in progress, but the white inclusions are undoubtedly the oldest surviving material found so far (FIG 2.7).

July 20, 1969 saw the successful landing of Apollo XI on the Moon. This event had been planned over the preceding decade and a large number of scientists of a variety of disciplines had during this time investigated terrestrial and meteoritic samples in preparation for receiving lunar rocks. The return of lunar samples was the moment of truth for a number of theories. There was evidence for extensive volcanism in the lunar 'seas', but no evidence that they had ever been wet (FIG 2.8). The lunar landings gave us a new insight into the earliest history of a planetary body which partially melted to form a crust, preserved in the lunar highlands, and which was bombarded by planetary debris during the first 600 million years of its history.

On 28 September, 1969, there occurred a second important meteorite fall from which some 500 kg of stones were collected near Murchison, in Victoria, Australia. In them was made the first undisputed discovery of amino acids of

FIG 2.8   Surface of lunar mare. View northwards over Mare Imbrium by the mapping camera on Apollo XV. The low angle of the Sun is ideal for revelation of topographic features. Note the lava flows (arrowed) extending on either side of the mare ridge that runs from upper left to lower right. The field was partially obscured on the right by the protective cover of the camera. NASA photograph, courtesy Lunar and Planetary Institute, Houston.

extra-terrestrial origin. Amino acids are the 'building-blocks' of protein; so investigation of the Murchison meteorite may have a direct bearing on our search for the origin of life. It must, however, be stressed that it is most unlikely that the compounds in Murchison were formed by living organisms.

The fourth important development in 1969 was the finding of meteorites on Antarctic ice. A Japanese expedition exploring part of Antarctica found nine pieces of meteorite, one of which was a new variety of a previously known, but rare, type. In the succeeding eight years Japanese and American expeditions, whose purpose was to recover meteorites, added about 5000 more to collections. Although many of the pieces represent fragments of a single fall (see Chapter 1), a number of rarities was encountered and, for the first time, it proved worth while to send expeditions to look for meteorites the falls of which had not been witnessed. While this account was being written, a Dane and an American were planning to visit north-east Greenland to try to repeat the successes of Antarctica, but unfortunately their plans failed to materialize.

Into the 1970s lunar landings continued. With six Apollo landings and three successful unmanned Soviet missions, we now have access to returned samples from nine locations on the near-side of the Moon. The Russians have landed several unmanned craft on Venus, two of which sent back pictures of a rocky surface. But most observations on Venus have centred on its dense atmosphere. American missions to Mercury (after passing close to Venus), Mars and Jupiter have greatly increased knowledge of our planetary neighbours. However, the highlight of the 1970s will probably prove to have been the landings of the two Viking craft on Mars in 1976. Not only did we receive pictures of two boulder-strewn, reddish landscapes, but also data on the chemical composition of the atmosphere and of the soil. Furthermore, the largely inconclusive experiments searching for evidence of life caused scientists to re-investigate the properties of some sterile soils from the Antarctic, to try to find an explanation for the martian results. In addition to the landings, the two parent spacecraft continue to orbit the planet, after having transmitted to Earth pictures of Mars and of its two misshapen, battered, moons. In 1979, the Voyager craft transmitted pictures of Jupiter and of some of its satellites. One satellite, Io, exhibited active volcanoes and Jupiter was found to possess a ring system, like those of its neighbours, Saturn and Uranus. Both space craft have visited Saturn and Voyager II is now on its way to Uranus.

Early in 1982, attention was again focused on Venus. A topographic map was prepared from data gathered from the radar on board Pioneer Venus Orbiter, launched by the USA in 1978, and the Russian landers Veneras 13 and 14 produced chemical analyses of near-surface materials.

For the future, plans are already being considered for obtaining returned martian samples, for an encounter with Halley's comet, and for missions to the asteroids. But before huge expense is incurred in bringing back a sample from an asteroid, we should first try to decide whether or not we now have a sample in the form of a costless meteorite. The last part of this chapter is devoted to an account of a new investigation of an old meteorite fall, that may link it to one of five asteroids.

We saw in Chapter 1 that all three meteorites, the orbits of which are precisely known from photographs (FIG. 1.3), came from the asteroid belt. Furthermore, the visual observational evidence from bright fireballs which produced meteorites also supports an asteroidal source for meteorites. It is thought that collisions between asteroids cause fragmentation; that some fragments of asteroids eventually travel in Earth-crossing orbits and so may become meteorites if swept up by Earth. Now, after a collision, the fragmented material normally follows an orbit closely similar to that of its parent asteroid. But with time the orbits tend to change because of the effect of gravity during close approaches to one or other of the neighbouring planets such as the Earth, or Mars, but especially the massive Jupiter.

Space is penetrated by cosmic radiation, which means electrically charged atoms travelling at close to the speed of light and in every direction. All but the most powerful cosmic radiation is stopped by a metre or so of stony material – or an atmosphere like the Earth's. Thus, when a potential meteorite is merely a portion of an asteroid, if buried a couple of metres below the surface it is exposed to virtually no cosmic radiation. But after the potential meteorite is broken from its parent-body, cosmic radiation produces small, measurable changes in the chemical composition. The charged particles of the cosmic rays interact with some of the atoms of the fragment in space to produce isotopes of the gases helium and neon. In addition, various other isotopes, including a radioactive aluminium isotope, are produced. From measurements of the abundances of helium and neon and, in favourable circumstances, of the radioactive aluminium, a 'cosmic radiation age' can be calculated for a stony meteorite. This is the time since the meteorite was freed from its parent-body until it landed on Earth; the time during which the potential meteorite was bombarded by cosmic rays. Most stony meteorites have cosmic radiation ages greater than some millions of years, which means that as small bodies the potential meteorites orbited the Sun during this time. But one meteorite, which fell in Kansas in 1890, has a cosmic radiation age of less than 25 000 years.

It has been calculated that *on average* a meteorite's orbit is changed once every 500 000 years by a close approach to a planet. This suggests that the average stony meteorite with a cosmic radiation age of, say, 10 million years, has had its orbit changed some twenty times. But its 25 000 year radiation age indicates that the orbit of the Farmington, Kansas, meteorite was close to that of its parent asteroid, because it is statistically improbable that the Farmington orbit was altered by a close approach to a planet. Thus, its fall to Earth was probably its first close approach to a planet. From this we may infer that the parent asteroid was also in an Earth-crossing orbit, which limits our choice of parent to one of the Apollo asteroids, a group of which only about a dozen bodies are known. If the orbit of Farmington could be reasonably established, it could lead to the identification of the asteroid from which it came. This achievement would eliminate at least one asteroid as the target for a space shot. But how can the orbit of a meteorite be deduced some eighty years after its fall? This challenge was taken up by an American and two Russians.

They engaged the help of the Historical Societies of Kansas and Nebraska. Issues of newspapers of towns adjacent to the place of fall were checked for days to weeks after the date of fall, 25 June, 1890. From 280 newspapers, 144 reports of the meteorite fall were obtained. However, many reports were duplicated, and the scientists finally selected 60 reports from 25 locations in and around the place of fall. They also received the report of a 98-year-old eye-witness!

The Farmington meteor must have been extremely spectacular, for, although it fell at 12.50 p.m. on a midsummer day, 'It was a wonderful sight and the light was fully as strong as the sun itself. It shot across the sky with the velocity of a rocket into midair.' From directional information, sometimes false, the three scientists deduced that the meteor most likely came from 20 degrees west of south. To some observers the meteor appeared to come from close to the Sun, which enabled the investigators to get a fairly accurate estimate of the angle of descent, about 60 degrees from the horizontal. The fireball probably extinguished about 10 kilometres (6 miles) up, after which two stones fell, the larger weighing 85 kg; much of these are preserved in the Field Museum of Natural History, Chicago. But there was no way of estimating the velocity of the meteor. In calculating the orbit, the scientists tried four different values for the velocity, the lower limit of 13 km per second being set by the Earth's escape velocity, the upper limit by the velocity, 22 km per second, above which the meteorite would have been completely burned up.

In spite of the fact that only an approximation of the Farmington orbit was obtained, it could be concluded that this meteorite is possibly related to one of three Earth-crossing asteroids, 1862 Apollo, Hermes, or 1865 Cerberus. Two others, 1620 Geographos and 1685 Toro are less likely candidates.

It is perhaps unfortunate that the results of this investigation were inconclusive, but it is a worthy illustration of the dependence of scientists on the non-scientific community for observations of unusual or transient phenomena.

# 3
# Fact and theory – our knowledge of the Earth, Moon and inner planets

All nine planets orbit the Sun in the same direction and in essentially the same plane. For convenience, the planets are often considered as forming two groups. The outer planets comprise Jupiter, Saturn, Uranus, Neptune and Pluto (FIG 1.3, p. 13). All but Pluto are larger and more massive than the Earth and are apparently composed of materials such as hydrogen, helium, methane and ammonia which would be gaseous or liquid under the conditions prevalent at the Earth's surface. Pluto seems out of place. With a radius of about 1800 km it is about the size of the Moon, but its low density of 0·5 to 1·0 indicates that Pluto is composed of ices, not rocks. The orbit of Pluto deviates considerably from circular and crosses the orbit of its neighbour, Neptune. The chance of collision is, however, remote, because Pluto's orbit is inclined at 17 degrees to the plane of the Earth's orbit, whereas Neptune's inclination is less than 2 degrees. Pluto and Neptune are in resonance, the former making two orbits in the time taken for Neptune to make three. This ensures that the planets never get close enough to each other to interact. Because Pluto is so different in size from its huge neighbours, it is often considered to be a displaced satellite of Neptune.

Jupiter is the heaviest planet, being more than three times the weight of Saturn and three hundred times heavier than Earth. Jupiter has at least 13 moons, Saturn has 17 and Uranus has five, in contrast to Mercury, Venus, Earth and Mars, which can only muster three moons between them. One puzzling feature about Jupiter and Saturn are their sources of energy, for they radiate away more energy than they receive from the Sun. Because the average density of the planets is not much greater than that of water, it seems impossible that radioactivity of a heavy element like uranium could be responsible. Let us hope that clues to this and other problems will be provided by new spacecraft in the not too distant future, for although the two Voyager craft produced a wealth of new information, more questions were posed than solved.

Mercury, the planet closest to the Sun, and its neighbours Venus, Earth and Mars, are grouped together as the 'terrestrial' or 'inner' planets. Of these, the largest and most massive is Earth. All four are composed of rock and metal, in contrast to the gaseous or icy outer planets.

The inner planet – Mars – and the outer planet – Jupiter – are separated by a distance of 5·5 million kilometres, or rather less than four times the average distance of the Earth from the Sun (Chapter 1). This great zone is occupied by the asteroids, small bodies the largest of which, Ceres, is only some 1,000 kilometres in diameter. The total weight of material orbiting the Sun between

Mars and Jupiter is estimated to be a tiny fraction of the weight of the Moon. So, on average, the asteroidal belt contains about 100 000 times less matter than the equivalent neighbouring planetary regions.

For the time being we shall neglect the empty zone of the asteroids and the cold regions of the outer solar system, and concentrate on the sun-warmed inner planets.

Because it is our home, the Earth is better known than any other planetary body. Even so, because of their inaccessibility, the inner depths of the Earth are largely unknown and for information we must rely on inferences based on physical measurements. For example, we *know* that parts of the solid Earth can melt, for we see the product coming out of volcanoes. But we do not know the precise mechanism by which a small proportion of liquid can separate from solid residue into a volume large enough to produce volcanic eruption. Nor do we *know* the processes that allow different types of lava to issue from a single volcano. Such processes must be *inferred* from the chemical compositions of the lavas and from what we *think* is the composition of the parent-rock, some tens of kilometres below the volcano.

However, the knowledge of the scientist is not as limited as it may at first appear, for various theories are based on such a large number of well-established facts that the theories must be essentially correct. Any alternative must be highly improbable. For example, no one has actually *seen* the generations of one species of plant or animal evolving into a new species, but the theory of natural selection in one form or another is supported by such a huge battery of observation that evolution is totally accepted by scientists. In this chapter is outlined the state of our knowledge – the facts – together with illustrations of the use of observation in theorizing.

The surface of the Earth comprises continents and oceans. In addition to the presence or absence of a covering of water, oceanic and continental areas differ markedly in the average composition of the underlying rocks. The large scale differences between continental and oceanic regions are due to the different kinds of igneous and metamorphic rocks typical of each. Continents are composed dominantly of granitic rocks (Plate 1a) which are rich in silicon, aluminium, sodium and potassium, and poor in iron and magnesium. Rocks beneath the oceans are typically basaltic, which means that they are rich in iron, magnesium and calcium. Basalts (Plate 1b) have less silicon and usually less sodium and potassium than granites. In both rock-types, oxygen is the most abundant chemical element, and geologists usually consider the other elements to be in combination with it. Table 3.1 illustrates simplified chemical compositions of basalt and granite. The higher iron oxide, magnesium oxide and calcium oxide contents in the basalt, at the expense of the oxides of silicon, sodium and potassium, mean that basalt is denser than granite. A cubic metre of basalt weighs around 2·9 tonnes compared to about 2·7 tonnes for granite. It is this property that accounts for the very existence of the continents, which

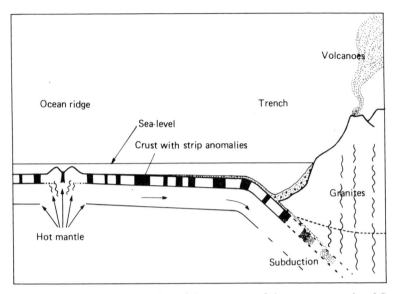

FIG 3.1   Schematic representation of the structure of the western margin of South America, not drawn to scale. Hot mantle rises, partially melts and produces basaltic rocks along an oceanic ridge (East Pacific Rise). As they cool the basaltic rocks become magnetized in the prevailing magnetic field (FIG 2.6). From the ridge the sub-oceanic crust is pushed apart, that to the east moving at a few centimetres per year towards the continent. As it does so it cools and contracts so the overlying ocean deepens. Sediments form on top of the crust. Near the continent the oceanic crust and part of the underlying mantle plunge under the continent – this is known as 'subduction'. The ocean floor is dragged down to form a trench that tends to fill with sediment eroded from the continent; some wet sediment is also subducted. The subducted mantle, oceanic crust and wet sediment are heated and partially melt to produce volcanoes and huge intrusive granite masses. The residual mantle, shed of its volatile and fusible constituents, becomes denser and continues to sink to a depth of at least 600 kilometres. The continent remains buoyant. Volatile-rich granitic crust overlies volatile-poor, denser lower crust of total thickness 40 kilometres or more, and much thicker than basaltic, oceanic crust.

apparently float on denser underlying rocks (FIG 3.1). But this is inference, not direct observation, and is based on the measurement of small variations in the Earth's gravity or of differences in the velocity of earthquake waves as they travel through the different layers of the crust. Chemical differences between rock-types are accompanied by different physical properties.

It is because of observed variations in physical properties within the Earth that we can infer that it is not uniform in chemical composition. Beneath about 30–60 kilometres under continents and about 10 kilometres below ocean-floors

TABLE 3.1
Chemical composition of granite and basalt

|  | Granite | Basalt |
|---|---|---|
| $SiO_2$ | 72·5 | 47·3 |
| $TiO_2$ | 0·3 | 1·0 |
| $Al_2O_3$ | 14.1 | 16.3 |
| $FeO*$ | 2·5 | 10.5 |
| $MgO$ | 0.5 | 10.2 |
| $CaO$ | 1·6 | 12.8 |
| $Na_2O$ | 3·3 | 1.8 |
| $K_2O$ | 5.2 | 0.05 |
| Sum | 100.0 | 99.95 |

The compositions are expressed in the usual geological way, as weight per cent oxides. *All iron is calculated to be in combination with oxygen in a 1:1 ratio. Granite and basalt always contain some water, but for comparative purposes the compositions are calculated water-free.

Si = silicon  O = oxygen  Ti = titanium  Al = aluminium
Fe = iron  Mg = magnesium  Ca = calcium  Na = sodium
K = potassium

Subscripts denote atomic ratios, e.g. $SiO_2$ means that each silicon atom is combined with two oxygen atoms; $Al_2O_3$, that 2 aluminium atoms are combined with 3 oxygens; no subscript means a 1:1 atomic ratio, 1 atom magnesium to 1 atom oxygen, etc.

earthquake waves travel faster than they do through basalt. This *suggests* that beneath both continents and oceans is a widespread rock-layer with a density of 3·3. From dredged samples, drill-cores and the accessible rocks of oceanic islands we *know* that in oceanic regions the uppermost layer of the solid Earth – the crust – is basaltic. We know too that continental crust is mainly granitic. But do we know with certainty the identity of the dense rock – the mantle – located below the crust? The answer is a qualified 'yes'.

In both continental and oceanic areas throughout the world certain basaltic rocks contain lumps of coarse grained, olivine-rich rock (Plate 1c). Laboratory experiments show that these 'olivine-nodules' have mineral assemblages stable at pressures from 10 000 to 20 000 atmospheres (10–20 kbar) and at temperatures in the range 1000–1400 °C. These are the conditions thought to prevail at depths between 30 and 60 kilometres below the surface. And since olivine-nodules occur world-wide they are thought to be from near the top of the upper mantle, the layer immediately below the crust. Similar rocks occur in large masses in mountain chains or in islands around the Pacific. Here, the olivine-rich rocks are thought to have been pushed up some tens of kilometres during mountain-building events.

But if olivine-nodules are stable at high temperatures and pressures, how do they manage to survive at the surface? The answer is that at high pressure and temperature the minor chemical elements in the rock are distributed differently among the four minerals present than they are at low pressures. If transport to the surface is rapid and accompanied by cooling, the minor elements are unable to migrate from their high pressure sites and enter their low-pressure host minerals. An 'unstable' mineral assemblage may thus be maintained indefinitely. For example, at pressures of 10 000 to 20 000 atmospheres and temperatures above 1000 °C, aluminium is distributed between three minerals, one being the dense oxide, spinel. In spinel, aluminium is combined with magnesium, iron and chromium, plus oxygen. At low pressure, aluminium, together with calcium and some sodium, is located almost exclusively in the silicate, plagioclase. Plagioclase has a density of 2·8 which is lower than that of spinel which approaches 4. Decompression of an olivine-nodule during transport to the surface requires the breakdown of dense spinel and the formation of less dense plagioclase. In practice, however, the complex mineralogical rearrangement seldom occurs.

In rare occurrences in very old parts of continents we are privileged to obtain samples of the Earth from a depth of perhaps 200 kilometres. The rock known as kimberlite is of economic importance as the primary source of the mineral diamond (Plate 1d). This rock is named after Kimberley in South Africa, where diamond was discovered in 1871. Kimberlite may be considered as a type of magnesium- and potassium-rich basalt with a high water content and is usually associated with calcium carbonate in veins. Under the great pressure obtaining at 100 to 200 kilometres below the surface, molten kimberlite has large amounts of water vapour and carbon dioxide gas dissolved in it. But if the molten kimberlite forces its way upwards to a high enough level, rupture of the overlying rocks may occur, usually along a deep-seated fissure. Pressure is immediately released, causing carbon dioxide and water vapour to come out of solution in the molten kimberlite, which froths upwards. Expansion of the gases (carbon dioxide and water vapour) causes very rapid cooling and the kimberlite is emplaced at or near the Earth's surface as blocks and debris of solidified rock deposited by the upward blast of gases.

Evidence of the cold emplacement of kimberlite is provided by the presence of diamond, which could not survive transport to the surface in hot, molten rock. Such rapid transit from depth is associated with abrasion of the surrounding rocks, and kimberlite usually contains blocks of rock torn from different levels. Laboratory studies indicate that some olivine-rich rocks, mostly of a type called garnet-peridotite, have mineral assemblages stable at 25 000 to over 50 000 atmospheres (25–50 kbar) and 800 to 1400 °C. We *know* from the experimental work that these rocks are stable under the conditions *estimated* to prevail at depths from 75 to about 200 kilometres beneath very old continental areas. Hence we are reasonably certain that we have access to samples from these depths in the Earth's upper mantle (Plate 2a).

In garnet-peridotite the mineral, spinel, is absent and most of the aluminium is contained in bright red, magnesium-rich garnet. The presence of this garnet is diagnostic of rocks of high-pressure regimes. In fact, it has proved to be very useful in diamond prospecting because it is much more abundant and more distinctive than the elusive diamond. When bright red, magnesium-rich garnets are found, there is a good indication that it is worth making a careful search for diamonds, with all the expense and effort that this will entail.

The range of chemical compositions found among olivine-nodules from shallow depths in the upper mantle is much the same as in the garnet-peridotite suite of rocks. Thus we have good information on the chemical composition of the Earth to a depth of 200 kilometres. This is important when we try to compare the Earth with the Moon and meteorites, a subject to which we shall return.

From our knowledge of the chemical variation in modern upper mantle we can obtain an estimate of its primordial composition. Earthquakes associated with volcanoes extend down to some tens of kilometres below the surface. This tells us that basaltic magmas are generated within the upper mantle. Laboratory studies of rocks thought to have come from the upper mantle show that some rocks can be partially melted to produce a small fraction of basaltic liquid. Most of the aluminium, calcium and sodium go into the liquid, leaving the solid residue impoverished in these elements, and when we look at rocks from the upper mantle we find that they have a wide range of aluminium, calcium and sodium contents. This observation is consistent with the idea that, over geological time, partial melting followed by basalt extraction has occurred to differing degrees in different parts of the upper mantle. Therefore, to estimate the composition of primordial upper mantle we must try to calculate the composition before any liquid was extracted. A good approximation may be readily obtained, because some garnet-peridotites and olivine-nodules do have compositions that could yield 10 per cent or more of basaltic liquid.

Table 3.2 illustrates the results of such calculations. The basalt, from the Isle of Skye in Scotland, was chosen because it is of a type that has low titanium and potassium. Such rocks are thought to have changed little in composition during their ascent to the surface. They are typical of the deep oceans, the ones on Skye having been erupted as part of the volcanism associated with the separation of Greenland from Europe some 60 million years ago. Upper mantle with the composition illustrated in the table can yield only 10 per cent of low-potassium basalt. This limit is set by the total removal of titanium. From the table, it is seen that the residue, after 10 per cent basalt extraction, is richer in magnesium but depleted in all the other elements relative to the starting upper mantle composition. We can conclude, then, that olivine-nodules or garnet-peridotites which are rich in titanium, aluminium, calcium, and so on, may have a composition close to that of primordial upper mantle. Upper mantle rocks that are depleted in these elements and enriched in magnesium have probably suffered at least one episode of heating with basalt extraction.

TABLE 3.2
*Chemical composition of upper mantle and basalt*

|  | Upper Mantle | Basalt | Residue |
|---|---|---|---|
| $SiO_2$ | 45·0 | 47·3 | 44·8 |
| $TiO_2$ | 0·1 | 1·0 | 0·0 |
| $Al_2O_3$ | 3·6 | 16·3 | 2·3 |
| $FeO^*$ | 8·0 | 10·5 | 7·5 |
| $MgO$ | 39·0 | 10·2 | 42·2 |
| $CaO$ | 3·3 | 12·8 | 2·3 |
| $Na_2O$ | 0·3 | 1·8 | 0·15 |
| $K_2O$ | 0.04 | 0·05 | 0·03 |
| Sum | 99·34 | 99·95 | 99·28 |

* All the iron is calculated as this oxide. The upper mantle composition is based on analyses of olivine-nodules from France and garnet-peridotites from South Africa. The basalt is specially selected for its low titanium content (see text). The residue is the calculated composition after 10 per cent of the basalt is removed from the upper mantle composition.

The upper mantle and residue have low totals because oxides of chromium and nickel are omitted for the sake of simplicity.

The approximation to primordial upper mantle composition outlined above works well for the more abundant chemical elements, but for many trace elements it fails. An element such as uranium has an atom so large that olivine, garnet and the common rock-forming minerals cannot readily accommodate it. In the upper mantle, uranium is largely concentrated along the boundaries between crystals or in small amounts of mica or other suitable hosts. This means that on heating to temperatures below that at which basaltic liquid could form and separate, uranium can migrate upwards through the upper mantle. It probably does so by transport in a water-bearing fluid moving along the inter-grain boundaries, but the mechanism is not properly understood. The net result, is, however, well known; granitic crust is enriched in uranium by 100 to 1000 times relative to modern upper mantle. There is just too much uranium in the continental crust for the uranium to be accounted for by extraction from the top 100 kilometres or so of upper mantle. Because of this, we cannot calculate accurately the uranium content of primordial upper mantle, and the same is true of other trace elements. They must have been supplied to the continental crust by the processing of an unknown depth of primordial mantle. To estimate the Earth's uranium content we must use indirect means such as first calculating the amount of heating thought to be due to the decay of uranium, thereby deriving the elemental abundance. With so many uncertainties, the Earth's uranium abundance is known to within about 30 per cent.

Our knowledge of upper mantle properties aids our understanding of the

Earth in another way. We know the size of the planet, the mean radius being 6371 kilometres. From the measurement of gravity we can calculate its weight. Combining these, we find that the Earth has an average density of over 5·5. Put another way, at the Earth's surface a cubic metre of 'average' Earth weighs 5·5 tonnes. Because the upper mantle density of 3·3 is less than this, we immediately *know* that part of the Earth must have a density greater than 5·5. Such evidence supports our interpretation that the paths of earthquake waves are refracted by a high density core (FIG. 2.5, p.32).

Seismic studies indicate that the core is in two parts. The inner core has a density of over 13, is probably solid and extends outwards to over 1,200 kilometres from the Earth's centre. The outer core extends from its junction with the inner core outwards to almost 3500 kilometres from the centre. Its inferred density ranges from 12 to 10 at the outside contact with the base of the mantle. The outer core acts as a liquid, for $S$, or shear, waves from earthquakes cannot pass through it. For $S$ waves the wave motion is at right angles to the direction in which the wave is moving. On the contrary, $P$, or pressure, waves do pass through the outer core. These are analogous to the ripples which radiate outwards when something is thrown into a pond. Together, the inner and outer core make up 32 per cent of the total weight of the Earth. The high densities are partly due to compression under the immense pressure obtaining at depth in our planet. When the densities are calculated for surface conditions it turns out that the core has a density just below that of pure iron, that of the outer core being lower still. On this basis it is usually argued that the liquid outer core is largely made of iron, but with up to about 20 per cent of some low-density element such as sulphur. Nowadays, by analogy with meteorites, sulphur is the most favoured element, but metallic silicon, phosphorus and oxygen have also been considered. By analogy with meteorites, too, six or seven per cent of nickel should be present in the solid inner core, and perhaps less in the outer core.

When oxidized iron in the mantle is added to the amount thought to be in the core, the Earth as a whole apparently has about 35 per cent of iron by weight. This is perhaps the best established piece of information we have on the bulk chemical composition of the Earth. But it is worth remembering that it is derived indirectly and other possibilities are considered less likely only because of what we know of meteorites and the outermost part of the Sun.

The Earth has a powerful magnetic field. When taken in conjunction with our information on the internal constitution of the Earth, this gives us vital clues in the study of other planetary bodies. It is thought that the Earth's magnetic field is related to motions in the outer core which produce a dynamo effect; that the Earth is a powerful electromagnet of variable intensity which can switch on or off at random and which can change its polarity. Effects of these magnetic changes have been 'frozen in' in various volcanic rocks erupted at different times over the previous tens of millions of years.

The presence or absence of a magnetic field in a planetary body can be established by a spacecraft. However, laboratory measurements on returned samples are required to obtain precise data on the past magnetic field. The size of a planetary body is measurable by optical telescope. Its gravity is now best measured by the effect it has on a near-by spacecraft. Together, these provide us with the mean density of the object. By analogy with the Earth, density and magnetic data allow us to estimate the iron content of the planetary body, the probable presence or absence of a core, its size if present, and whether it is likely to be liquid or solid.

In addition to knowing the chemical compositions of the rocks of our continents and oceans, we have direct evidence for their ages. We find that the oldest rocks of the ocean floor are a mere 160 million years old. These are the oceanic rocks closest to continental margins. The youngest oceanic crust is still being generated along the ridge systems of the oceans, for example, in Iceland.

For the continents the story is different. Some parts of the continental crust have remained unchanged for at least 3700 million years. Around these, our most ancient rocks on Earth, are zones of younger rocks some 2800 million years old that are surrounded in turn by still younger rocks. This much we *know*. The *inference* is that granitic, continental crust was formed at different times but, once formed it was not destroyed. If continental crust is not destroyed, the continents have been slowly growing throughout geological time. Now, laboratory melting experiments on upper mantle rocks always produce basaltic liquids. It therefore seems unlikely that new, granitic crust is produced directly by partial melting of the upper mantle. Instead, a two-stage process appears to be required. The upper mantle partially melts to produce a small fraction of basaltic crust. Heating of the base of the basaltic crust may then produce a small fraction of granitic liquid that migrates upwards to solidify as new, continental crust, never to be returned to the mantle. Further addition of trace elements may occur after they have been 'sweated out' of upper mantle or from basaltic material immediately underlying the granitic crust.

It is amazing that, over the past 4000 million years, oceanic crust may have been formed and destroyed a number of times. We are lucky that the continents are stable, for stability of environment over long periods must have been essential for the evolution of the variety of life forms on Earth.

So much for the broad features of the Earth; let us now turn to our only natural satellite. Unlike the Earth, the Moon has no atmosphere. Gravity on the lunar surface is only one sixth that on Earth and is not strong enough to hold any gases that might be evolved from within. The Moon has a radius about a quarter that of the Earth and its weight is less than one eightieth of the Earth's. But compared with the satellites of the outer planets the Moon is fairly large, and, *relative to the weight of its companion planet* the Moon is the heaviest satellite in the solar system. Titan, the largest of Saturn's moons, is less than one thousandth of Saturn's weight; and Ganymede, the largest Jovian satellite, is less than one ten-thousandth of Jupiter's weight. This is worth remembering when we come to discuss the origin of the Moon, for the relatively tiny satellites of the outer planets may have originated quite differently.

FIG 3.2 Sites on the Moon from which samples have been returned.
XI, Apollo XI, Mare Tranquillitatis – July 1969.
XII, Apollo XII, Oceanus Procellarum – November 1969.
XIV, Apollo XIV, Fra Mauro – January 1971.
XV, Apollo XV, Hadley-Apennines – July 1971.
XVI, Apollo XVI, Descartes area – April 1972.
XVII, Apollo XVII, Taurus-Littrow – December 1972.
16, Luna 16, Mare Fecunditatis – September 1970.
20, Luna 20, Apollonius highlands – February 1972.
24, Luna 24, Mare Crisium – August 1976.
(Photograph of the Moon, courtesy of Lick Observatory)

Observation from Earth established that the Moon has reflective, bright, mountainous areas and dark, low-lying areas, the outlines of which may be curved. Distributed overall are craters of variable size; some are very bright and central to radiating bright rays. In general, there are fewer craters per unit area in the dark terrains than on the light-coloured highlands. But only just over half the Moon's surface may be viewed from Earth and until information was received from orbiting craft, the lunar far side was a closed book. To our surprise it was discovered that the far side consists almost entirely of bright highlands. Lunar features are not symmetrically distributed over the surface.

This was the extent of our knowledge until lunar samples were returned, and when this occurred a new era of discovery was immediately opened. Rocks from six manned Apollo flights and three unmanned Russian missions are now available for laboratory study (FIG 3.2). However, the huge success of Apollo was not based on returned samples alone, but also on the number and variety of physical measurements carried out. These provided information on the internal structure of the Moon, on the chemical composition of the surface and the uranium content of the interior. The flow of physical data has now ceased, but work continues on the returned samples.

Rocks from the dark, low-lying areas – the maria – are volcanic in origin and not unlike the basalts of Earth. Light-coloured calcium and aluminium-rich rocks are prevalent in the lunar highlands. But because of the effects of meteorite impact, both rock-types occur in differing proportions in all sites sampled. Projectiles hitting the Moon are not slowed down by an atmosphere, as on Earth, and even the smallest strikes the surface at a velocity of some kilometres per second. Fragments are showered far and wide, and a small proportion of the surface materials is completely melted. In almost every case the projectile is vaporized and distributed throughout the comminuted target. Over millions of years this produces lunar 'soils' with components from events widely spaced in time and place. Impact has been so efficient in covering the surface that no bedrock has yet been sampled.

Let us now look more closely at the rocks of the lunar highlands and maria. We have no direct evidence, as we have for the Earth, for the identity and composition of sub-surface lunar rocks, so theories of the composition of the Moon depend largely on the assumptions and preferences of the theorist. We will try to follow the same sequence of reasoning that we used with the Earth, beginning with fact and then attempting to form a coherent picture by interpreting physical data in terms of what we know.

The denser cratering of the lunar highlands indicates that they are older than the maria. In the Apollo XI sample, coarse-grained rocks rich in the calcium aluminium silicate mineral anorthite were identified as being highland-derived. (Anorthite is an almost sodium-free plagioclase, and of low density.) Subsequent missions added olivine-bearing and other types to the inventory of highlands rocks (Plate 2b). Even before the Apollo missions, it had been noted that around the margins of some maria, highlands rocks and structures appear to have been swamped by mare materials (FIG 3.3). It then follows that the mare basins are underlain by highlands rocks, which presumably have a moon-wide occurrence, and represent pre-mare lunar crust. This is of great importance for our understanding of the early history of the Moon and its discovery has led to a reappraisal of early, crust-forming processes on Earth.

FIG 3.3   Mare Imbrium on the Moon. A circular basin excavated some hundreds of millions of years before it was largely filled with lavas. At A, a ridge concentric with the mare margin has been breached by lava (smooth) from the mare. The crater, Plato (B), formed after the excavation of the circular basin, but before the lava outpouring, which partially filled it. In contrast, compare the fresh, prominent crater (C), named Aristillus, which post-dates eruption of the mare lavas. Note the denser cratering in the ancient highlands terrain compared to the younger mare surface.

Photo courtesy of the Mount Wilson and Palomar observatories.

The rocks of the maria are dark and dense. They are termed 'basaltic' but differ from terrestrial basalts in having high iron and low sodium and potassium contents (Table 3.3). Although there was great excitement when it was found

TABLE 3.3

*Chemical composition of anorthosite, mare basalt and terrestrial basalt*

|  | Lunar anorthosite | Mare basalt | Skye basalt |
|---|---|---|---|
| $SiO_2$ | 44·4 | 40·7 | 47·3 |
| $TiO_2$ | 0·2 | 10·2 | 1·0 |
| $Al_2O_3$ | 29·5 | 10·5 | 16·3 |
| FeO | 3·6 | 18·8 | 10·5 |
| MgO | 5·0 | 7·0 | 10·2 |
| CaO | 16·6 | 11·7 | 12·8 |
| $Na_2O$ | 0·7 | 0·4 | 1·8 |
| $K_2O$ | 0·06 | 0·1 | 0·05 |
| Sum | 100·06 | 99·4* | 99·95 |

*The total is low because small amounts of chromium, sulphur and other elements have been omitted.

Note the extreme enrichment of the lunar anorthosite in aluminium and calcium. The mare basalt is a titanium-rich type; note its higher iron and lower magnesium relative to the basalt from Skye, but the latter has the higher sodium content.

that the basalts of Mare Tranquilitatus (Apollo XI) are titanium-rich, subsequent missions brought us basalts with much lower values. The Moon may thus not be much richer than the Earth in this element. By far the most significant feature of the mare basalts is their high oxidized iron content, which gives the rocks a density of 3·3. Furthermore, laboratory experiments have shown that a high iron content renders a basaltic liquid extremely fluid. As a result mare basalts form flows only a metre or so in thickness but covering areas of hundreds of square kilometres. In the maria, successions of such flows tended to have horizontal surfaces closely following the contours of pre-existing features, and would in this respect resemble the oceans on Earth. But the Moon was completely dry. Lack of water and oxygen in the lunar environment have meant that the rocks which survived damage by impact are in the same condition today as they were when extruded thousands of millions of years ago.

Although it had long been known that the lunar highlands are older than the maria, how much older was indeterminable until results were obtained from returned samples. Rocks from the highlands were found to range from 4200 to 3900 million years in age and up to 1000 million years older than the mare basalts, which have ages of 3800 to 3200 million years. Exceptionally, a few highlands rocks have ages exceeding 4200 million years but, at the other end of the time-scale, no crystalline mare rock studied so far is significantly younger than 3200 million years. Thus the lunar maria had formed by that time. Heating within the Moon must therefore have been short-lived, lunar volcanism having ceased before most of the Earth's continental crust had evolved. But when did lunar heating begin? We have no precise answer. One rock from the highlands crystallized over 4500 million years ago, but most of our information merely tells us that there were two, or more, heating events between 4600 and 4200 million years ago. All we can be sure about is that highlands-type crust formed in this time-interval during a period of intense meteorite bombardment. At the end of this time occurred either the last major impacts or else a brief period of abnormally large impacts. These created the huge basins that were later filled, or partially filled, with basalts to become the maria we know today.

Utilization of physical measurement adds detail to our sketchy picture of lunar history. Data transmitted to Earth from seismometers on the lunar surface indicate that the low-density, highland-type crust extends down to about 55 or 60 kilometres. Beneath this, faster seismic velocities require the presence of higher-density material down to about 750 kilometres below the surface of the Moon. These rocks may be similar in composition to the olivine-rich upper mantle on Earth. At depths greater than 750 kilometres the Moon is probably partially molten. When a quake or meteorite impact occurs on the far side, $S$ waves are not always received by every seismometer. As for the Earth's outer core, the interpretation is that the deep interior of the Moon may be partially molten. Finally, it is probable that the Moon has a small metal-rich core up to about 350 kilometres in radius and constituting no more than 5 per cent of the total weight of the satellite.

Even before manned lunar flights it was found that orbiting craft were accelerated towards circular mare areas and retarded during departure from over them. This can be due only to a local increase in gravity near the mare surfaces. A gravity increase, in turn, implies the presence of rocks of higher than average density. Concentrations of weight, or mass, in the mare regions have been termed 'mascons'. Their interpretation is by no means without dispute, but all agree that dense, mare material must be supported by an underlying, rigid, less dense layer – presumably the early lunar crust. Thus, lunar crust must have been rigid and, therefore, cold some 3800 million years ago when the first mare basalts were extruded on to it. Otherwise they would have sunk into the crust until stability had been obtained, and stability would not have produced a gravity anomaly.

Today, the Moon has a very weak magnetic field, so that it cannot now have an active, molten core like the 'dynamo' thought to exist around the centre of the Earth. But in the distant past there may have been a molten lunar core. Mare basalts *are* magnetic. They were magnetized by cooling in a magnetic field. If a magnetic field is present when certain iron-bearing minerals cool through particular temperature ranges, the minerals become magnetic. If the field dies away, the minerals still retain their magnetism. The strength of their magnetism depends not only on their cooling history but also on the strength of the primary field which induced the magnetization. Laboratory measurement of the magnetism existing in mare basalts shows a tendency for the older basalts to be more strongly magnetized than the younger ones. Therefore it is suggested that a magnetic field of declining intensity was present during the eruption and cooling of mare basalts (FIG 3.4). The question is, was the field produced by a lunar dynamo? Other possibilities are that the field was induced by electric currents caused by heating and friction during impacts, or by an intense solar wind. Professor Runcorn champions the dynamo theory. If he is correct, motions in a small lunar core steadily declined from 3900 to 3200 million years ago, after which it solidified. But we know from the presence of mascons that the lunar crust was rigid and cold during this period. Therefore, argues Runcorn, the metallic lunar core must have had a source of heat not present in the stony crust and mantle. Furthermore, the heat-source had become extinct by around 3000 million years ago. Presumably, the heating was caused by the energy released by decay of a radioactive element. But no suitable radioactive element known has an affinity for metallic iron or iron sulphide, the likely materials of a lunar core. So an as yet undiscovered super-heavy element was suggested. Physical theory indicates that beyond the natural and artificial elements (including uranium and plutonium) atomic stability will again be achieved in atoms of much heavier weight. These could have been produced in the same event as some of our calcium, iron and uranium, and perhaps only 100 million years before the solar system formed. The search for a super-heavy element continues.

Fact and theory may be combined to produce the following synthesis of lunar history. The Moon formed at least 4500 million years ago. Within a few hundred million years the upper part, or all, of the body had melted. Low-density, anorthite-rich material floated to form a layer some 60 kilometers thick which had completely solidified by 3900 million years ago. Below the crust the remainder of the early melt crystallized to form denser iron and magnesium-rich rocks. During this period the Moon was heavily bombarded and large depressions, often circular, were blown in the early crust. Partial melting occurred beneath the depressions. Either the dense fraction below the crust was remelted, or else deeper levels may have been melted for the first time. In either case, the result was the flooding of the basins by mare basalts until activity ceased at around 3200 million years ago. Since then sporadic impacts continued, forming 'soils' and producing the bright-rayed craters.

But even this simple scenario is not without its problems. Build-up of heat from decay of long-lived radioactive elements such as uranium probably caused the *second* period of partial melting associated with mare volcanism. But what caused the first? We do not know. Although highlands rocks contain the diagnostic imprint of the decay products of plutonium, they are not in sufficient quantity for the fission of plutonium to have caused melting. As we shall see, there is evidence for strong, early heating in the small parent-bodies of meteorites and the identification of the heat-source is one of the main problems in planetary science. However, given that bodies of asteroidal size melted and that their interiors stayed hot for millions of years, the Moon may have been heated by a combination of residual heat and gravitational energy from the accretion of

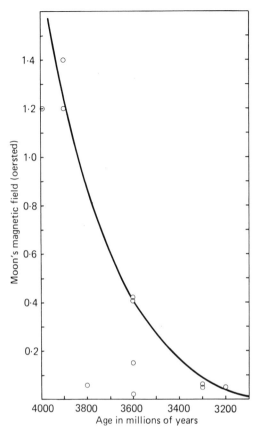

FIG 3.4   Decline of the Moon's magnetic field from 4000 to 3200 million years ago. The measurements indicate the ambient magnetic field necessary to produce the magnetism observed in the rocks today when the rocks cooled from high temperature. Spurious, low measurements may be due to loss of some magnetization during reheating by impacts. More recent work tends to support the curve, which indicates that the Moon's field halved in intensity roughly every hundred million years.

Courtesy Dr A. Stephenson, University of Newcastle upon Tyne.

such bodies – see Chapter 7. The Earth is still active; the Moon became dead some 3200 million years ago; Mars may be an intermediate case.

Mars is well known for its red colour and from the evidence that it is *the* planetary neighbour most likely to sustain life. The belief that organized life existed on Mars stemmed largely from the telescopic observations of 'canals', but unfortunately visiting spacecraft in the 1960s and '70s have destroyed the illusions of the telescope; there are no canals on Mars. As a consolation, these same craft have shown us that the red planet has surface features on a very large scale and these are confidently interpreted as volcanoes and systems of canyons. Telescopic observation from Earth proved that Mars has an atmosphere, albeit thin, and indicated that there are polar ice-caps; in this case the spacecraft have confirmed the earlier Earth-based interpretation. But because the planet Mars is never closer to Earth than 60 million kilometres (40 million miles), Earth-based observations could not compete with imaging by spacecraft for details of the Martian surface. However, we had to wait until the Mariner 9 craft had been orbiting Mars for a month or more before we had a proper glimpse of the variety of features on the planet. As a result of recent observations, and especially those of Vikings I and II, our knowledge of Mars has increased enormously. However, we still have few 'hard' facts about the planet and so care must be taken in interpreting the meagre data we do have. In this context we ought to remember some of the misinterpretations of lunar data before Apollo XI.

The average radius of Mars is slightly less than 3400 kilometres, or a little more than half of Earth, and less than twice that of the Moon. Mars weighs just over a tenth of the weight of the Earth, and gravity on the Martian surface is barely two-fifths that on Earth. The mean density, 3·9, suggests that Mars could have a metallic core constituting up to 19 per cent of its total weight. However, there is no detectable magnetic field and if a core does exist it cannot be molten and in motion, for the planet rotates at much the same speed as Earth and so a dynamo effect could be expected. At the surface, the thin Martian atmosphere exerts a pressure less than one hundredth that on Earth. Carbon dioxide is the most abundant gas, followed by nitrogen, argon, water vapour and oxygen. Solar heating produces wind speeds estimated at over 200 kilometres (130 miles) per hour and capable of carrying dust to a height of at least 20 kilometres. This concludes our list of physical data, but much information on planetary processes may be gained from observations of land-forms and from estimates of the chemical composition of the soil at the two Viking lander sites.

The first three missions to Mars were 'flybys' and in total only one tenth of the planet's surface was imaged. Moreover, the pictures showed variously cratered areas not unlike the different parts of the Moon. Then, during November, 1971, Mariner 9 became the first craft from Earth to orbit Mars. But the whole planet was obscured by dust, with only a few 'spots' showing through. To the delight and surprise of the watching scientists, when the dust began to settle, the 'spots' proved to be the tops of huge volcanoes on a much larger scale than anything on Earth (FIG 3.5). One of them, Olympus Mons, is about 22 kilometres high and its base is 600 kilometres across. For comparison, Mount Everest is less than 9 kilometres above sea-level. In addition, unlike the rocks of the lunar maria, some Martian volcanoes are almost free from impact craters. Martian volcanoes are therefore younger than their lunar counterparts because the more massive Mars should receive an even greater amount of debris from space than the Moon. Mars may still be capable of volcanic eruption; if not, volcanoes were active in the not too distant past.

FIG 3.5   A Martian volcano. Olympus Mons is an immense volcanic edifice some 22 kilometres high and 500 kilometres across. The summit caldera (subsidence crater) is 65 kilometres wide (between arrows). Around the volcano is an escarpment 2 kilometres high (arrowed, on right) and of unknown origin. NASA mosaic, courtesy Lunar and Planetary Institute, Houston.

Canyons and river-like channels are enigmatic features of Martian topography, the question being: 'Was liquid water involved?' Like the volcanoes, several canyon systems are larger than their terrestrial equivalents. The canyons of Mars appear to have been fed by branching tributaries and some canyons show 'braiding' analogous to the structure of a river moving across a flood-plain on Earth. Analogy, however, is not the same as proof, and the Martian canyons may have been sculpted by a fluid other than water. But the evidence for wind erosion and transport seems indisputable, and various dunefields are known.

The most recent success in the saga of United States exploration of Mars was the soft landing of two automatic craft in July and September, 1976. Vikings I and II were put down on low-lying plains areas about 6500 kilometres (over

4000 miles) apart. In each location the landscape was boulder-strewn and red in colour (Plate 2c). In addition to a package of three experiments designed to test for evidence of life – the results were inconclusive – the two craft also carried an instrument for chemically analysing the 'soil'. Two radioactive sources were used, one at a time, to irradiate with energetic X-rays a sample scooped from near by. Input of X-ray energy excites the atoms in the sample and causes them to emit their own X-rays. The emitted X-rays have energies characteristic of the chemical elements emitting them. Measurement of the energies and quantities of emitted X-rays tells us what elements are present and in what proportions. By laboratory standards, the results were imprecise, but from an instrument weighing only 2·1 kg (4·5 lb) sitting on a planetary surface some hundreds of millions of kilometres from Earth, they were impressive. Both soils from widely spaced localities have similar compositions. Expressed as oxides as in the previous tables, the average composition is approximately as follows: $SiO_2$ 45; $TiO_2$ 0·9; $Al_2O_3$ 5·7; FeO 16; MgO 8; CaO 6; all are in weight per cent. In addition, 3 per cent sulphur was measured.

Although the aluminium is low, this composition is not unlike that of some lunar basalts, but in the oxidizing environment of Mars the iron probably occurs in the more oxidized, ferric form, $Fe_2O_3$. In fact, the overall red coloration at the landing sites could well be due to a high rust content. However, rust cannot be present as a thin coating covering all grains. For if this were so, the low energy X-rays from magnesium, aluminium and silicon would have been strongly absorbed by the iron in the rust, even if the coating were less than one thousandth of a millimetre thick. Such a coating would have rendered the three elements undetectable in the X-ray experiment. Perhaps a more likely explanation of the red colour is the abundance of iron-rich clay. Laboratory experiments and calculations indicate that under the conditions prevailing on the Martian surface, igneous rocks would tend to decompose to clays, and that the planet could have a coating of clays up to a metre, or more, in average thickness. The high sulphur abundance is worth comment, as it is about one hundred times greater than could be expected in a soil on Earth. But the presence of sulphur in an oxidized environment is entirely consistent with the observation of a hard crust on the Martian soil. This *may* be due to sulphates of calcium (the mineral, gypsum?) or magnesium (the mineral epsom salt, or epsomite?) forming a 'hard-pan' as in desert areas on Earth. Here, seeping water deposits compounds in a layer close to the surface and thus forming a hard cake. But for the moment, this explanation for Mars must be purely speculative. To answer a number of questions about Mars, we need wait for samples returned to Earth.

Mars has two small moons, Phobos (Fear) and Deimos (Flight), named after the Greek mythological horses that pulled the war god's chariot. Phobos, the inner moon, revolves around its planet over three times in a Martian day of 24·5 hours. The orbit of Phobos is nearly circular and less than 6100 kilometres from the surface of Mars, which is close to the distance at which gravitational break-up of an unconsolidated body should occur. The satelite has an irregular

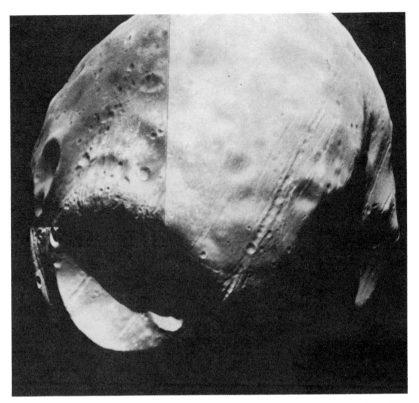

FIG 3.6   Phobos, the inner moon of Mars. An irregularly shaped, battered satellite. The parallel striae (see text) are clearly visible, trending from upper right to lower left. NASA mosaic, courtesy Lunar and Planetary Institute, Houston.

shape and a heavily cratered surface on which are parallel striae (FIG 3·6). These may be due to incipient fracturing caused by gravitational effects. We know that the maximum diameter of Phobos is only 27 kilometres, but because its shape is irregular, the volume is imprecisely determined and so estimates of the density are only approximate. However, 2·2 is the figure now generally accepted. This density is too low for rocks such as basalt and suggests that Phobos may be composed of water-bearing material, like the meteorites known as carbonaceous chondrites. Deimos has a maximum diameter of only 15 kilometres. It, too, is misshapen and heavily cratered. Both moons are more densely cratered than their planetary companion, suggesting that it may have had a different history; in fact, the moons of Mars may be captured asteroids.

Venus is the planet which approaches Earth most closely, the distance between them occasionally being as little as 40 million kilometres (25 million miles). In spite of this, we have less information about Venus than for Mars, for the former is shrouded in a dense, hot atmosphere rendering the surface invisible from Earth. In size, weight and density, Venus closely resembles Earth, each of these parameters being slightly less than its terrestrial equivalent. Venus probably has a metallic core, but has no detectable magnetic field. The reason could be either that the core is not molten, or, if molten, that the rotation of the planet is too slow to produce electric currents by friction between core and mantle, as probably occurs on Earth. The Venusian day lasts 243 Earth days! And on Venus, because the slow rotation is in the opposite direction from that of Earth, the sun rises in the west.

The atmosphere is dominantly composed of carbon dioxide, and high-level clouds have properties matching those of droplets of sulphuric acid. The clouds reflect four-fifths of the sunlight falling on them, rendering the planet visually bright. They also obscure the surface. Radar, however, can 'see' through the clouds and Earth-based radar was used to determine the broad features of the surface and hence the period of rotation. Since 1978 more precise topographic measurements have been carried out by the radar of the orbiting US spacecraft, Pioneer Venus Orbiter, which has now mapped over 90 per cent of the planet's surface. The polar areas remain unmapped because they are inaccessible to the orbiting radar by virtue of the craft's orbit. On Earth, the abyssal plains some 5 km below sea-level account for over half of the planetary surface, the bulk of the remainder being continental and above sea-level. In contrast, the surface of Venus is of uniform altitude, except for a very small proportion of high-lying terrain confined largely to three areas. Although some of the landforms in the three highland areas are reminiscent of volcanic or large-scale tectonic features on Earth – troughs flanked by mountains, for example – the lack of overall pattern suggests that plate tectonics have not shaped the modern Venusian surface. If this is so, then how does the planet get rid of its excess heat? The suggestion is that, again unlike Earth where most internal heat vents in volcanism along plate margins, Venus releases its heat through a few major volcanic areas. These have been identified by the presence of volcanic topographic features and by a radar 'signature' similar to that of lava flows on Earth. In addition, lightning which occurs in the atmosphere of Venus is concentrated over two 'volcanic' areas suggesting that they may still be active. (Lightning is often associated with volcanic eruption on Earth.) Other relevant information had already been supplied by Russian spacecraft.

In 1975, the Soviet Venera 9 craft transmitted an image of a surface strewn with bouldery rubble (FIG 3.7). When the craft landed its position relative to major surface features was unknown, but the mapping by the Pioneer Orbiter since showed that Venera 9 landed on the north-east side of Beta Regio, one of two highland areas considered to be volcanic. The blocks of rock in the Venera 9 image are angular, suggesting that they are fresh and unweathered. In contrast,

FIG 3.7   Bouldery rubble on the surface of Venus 'seen' from Venera 9. The location is to the east of Beta Regio, a prime candidate for an upland volcanic area which may be active still. Note the angular blocks (one is arrowed) that are consistent with a recent, volcanic origin. Old rocks would tend to be eroded and hence rounded in shape. Courtesy Popperfoto.

rocks on a surface south of Beta Regio (Venera 10), on rolling plains still further south (Venera 13), or in a basin to the south-east (Venera 14), are platy and apparently represent material that has been weathered and reworked into sediments. The Russian craft measured surface temperatures close to 450°C and pressures of about 90 atmospheres. In such a hostile environment it is not surprising that all craft have failed within a few hours of landing, and some even failed during descent. The USSR is therefore to be complimented on its successes which include measurements of the radioactivity at three sites and chemical analyses of near-surface materials at a further two sites. By 1975 we had learned that radioactivity on the Venusian surface is similar to that on Earth, indicative that Venus has partially melted to produce both granite-like and basalt-like materials. The most recent results take this conclusion further.

In March 1982, Veneras 13 and 14 analysed samples drilled from the surface. Analyses were performed by an X-ray technique not unlike that employed by the Viking landers on Mars. Venera 13 landed on rolling plains high on the south side of Beta Regio. Pioneer Venus Orbiter had shown this area to be an uplifted dome with a valley running across it. Such a structure may be analogous to East Africa, which is uplifted, with the Rift Valley running roughly north-south through it, as a zone of active volcanism. On Earth, volcanoes of continental rift valleys tend to produce lavas that are rich in sodium and potassium, and in titanium. Although sodium could not be determined by the Venera instruments, the Venera 13 sample was found to be rich in potassium and fairly rich in titanium (Table 3.4). Collaboration between the scientists of the USA and USSR enabled the selection of a site of different topography for the landing of Venera 14. This was in a low-lying basin to the south-east of Beta Regio. Material from this site proved to be poor in potassium and titanium and akin to basalts of the deep ocean basins on Earth (Table 3.4). On Earth, however, the eruption of these low-titanium and low-potassium basalts is linked to plate tectonics and typically occurs along mid-ocean ridges. It may, therefore, be

TABLE 3.4
*Chemical composition of the Venusian surface and terrestrial basalt*

|  | Venera 13 | Venera 14 | Basalt |
|---|---|---|---|
| $SiO_2$ | 45 | 49 | 47·3 |
| $TiO_2$ | 1·5 | 1·2 | 1·0 |
| $Al_2O_3$ | 16 | 18 | 16·3 |
| FeO | 9 | 9 | 10·5 |
| MgO | 10 | 8 | 10·2 |
| CaO | 7 | 10 | 12·8 |
| $Na_2O$ | – | – | 1·8 |
| $K_2O$ | 4 | 0·2 | 0·05 |
| Sum | 92·5 | 95·4 | 99·95 |

premature to dismiss the possibility that plate tectonics played a part in the development of Venus.

It may at first seem surprising that Venus, although matching Earth in size and density, is, in contrast, totally inhospitable. The explanation, however, is probably simple. Because Venus is closer to the Sun than Earth, the temperature of the Venusian surface must always have been higher. If radioactive heat-production on Venus is similar to that on Earth, the hotter Venusian surface would allow less internal heat to be lost. Hence temperatures below the Venusian surface would always have been greater than on Earth. In fact temperatures on Venus were probably so high that nowhere on the planet were carbonate minerals (such as calcite or dolomite) or hydrous minerals (such as clays and micas) stable. This means that on Venus there was no sub-surface or surface reservoir for carbon dioxide or water, which may be held in terrestrial rocks down to depths of 120 kilometres or more. The result is that Venus has a dense atmosphere rich in carbon dioxide. This gas admits sunlight to the surface, where the energy is absorbed and re-emitted at infra-red wavelengths. However, carbon dioxide, with minor amounts of water and sulphur dioxide, is opaque to the longer wavelength, infra-red rays and so the small proportion of solar energy that reaches the surface tends to be retained by the planet as heat. This is known as the 'greenhouse' effect. The high surface temperature so produced means that, as radioactive heating has declined, internal temperatures are still kept above the limit of stability of carbonate minerals.

Nearer to the Sun than Venus is the planet Mercury. It is a small body of radius only 2440 kilometres and not much larger than the Moon. Mercury weighs just over one-twentieth of the weight of the Earth. But Mercury's density, 5·45, is almost as high as the Earth's and because compression in the interior is low it is implied that Mercury has a high metallic iron content. To be precise, Mercury could have a core with a radius three-quarters that of the

FIG 3.8    Mercury from Mariner 10. North is at the top. Note the abundance of impact craters, many with bright rays. Part of a large, multi-ringed (impact) basin is visible on the left (arrowed); its diameter is 1300 kilometres. The image was taken when the craft was 210 000 kilometres from the planet on 29 March, 1974. NASA mosaic, courtesy Lunar and Planetary Institute, Houston.

planet and making up some two-thirds of its total weight. We might, therefore, expect Mercury to have a magnetic field and, indeed, this is the case, but its intensity is only one-hundredth of the Earth's field.

Discovery of Mercury's magnetism was made by Mariner 10 which was sent close to Venus, on past Mercury in March 1974 and finally into solar orbit. During two more close approaches to the innermost planet the craft sent back even more information. No significant atmosphere is present and the planetary surface was shown to be heavily cratered and not unlike the lunar highlands in appearance (FIG 3.8). Craters range from young, bright-rayed types to old, more subdued scars. Although at least one circular basin is present there is apparently no equivalent of lunar maria on Mercury. A second difference with the Moon is that some large-scale features on Mercury appear to have been caused by compression. These may have been produced by contraction of a large metallic core. Stony minerals contract less than metal, so shrinkage of a central core would set up compressive stresses within a thin, brittle, overlying silicate mantle.

In summary, we can say that from Mars to Mercury impact cratering is the rule. It must then be assumed that Earth has received its share of impacting objects, but that the destruction so produced has largely healed because the Earth is still active. The Moon and Mercury are now inactive bodies; Mars may be active still; but we can say nothing of Venus. Earth and Mercury have magnetic fields indicating the presence of metallic cores. Venus is almost as dense as the Earth and may have a core also. Mars and the Moon are of lower density, indicative of a lower ratio of metallic iron to stony minerals in them than in the three other planets. Mars could have a core up to almost one fifth the planetary weight; a lunar core seems less probable, the upper limit to its weight being one twentieth of the total.

From our discussion we can conclude that the inner planets and their satellites differ one from another in size, weight, chemistry and history. Possible causes may be inferred from the study of meteorites.

# 4
# Meteorite diversity

In this chapter the characteristics of the different meteorite groups are outlined. Because the constituent minerals and chemical compositions of many meteorites have remained unchanged for over 4500 million years, meteorites can tell us about *some* of the materials available at the time of construction of the planets. However, because the Earth, Moon and, almost certainly Mars, have been active during long parts of their histories, a direct comparison between their surface rocks and most types of meteorites is not strictly valid. For although our ideas of the composition of these planetary bodies as a whole are based on what we know about meteorites, the processes linking the two are largely speculative. But first, let us be factual – speculation can wait until Chapter 5.

Meteorites are broadly divided into three major categories: stony meteorites, iron meteorites, and a group with roughly 50:50 mixtures of stony and iron minerals called, not surprisingly, the stony-irons. Stony meteorites – usually just called 'stones' – are by far the most abundant of the three types. In meteorite research it is usual to distinguish between meteorites recovered soon after their fall was observed – the 'falls' – and those found on Earth after an indeterminate time of residence – the 'finds'. This distinction may prove important because finds are generally more likely to be contaminated with terrestrial substances introduced during rusting than are observed falls.

Most irons and stony-irons are finds (Table 4.1) and it was once considered possible that in the past the influx of stony meteorites was less than at present, or else the influx of irons and stony-irons was greater. In recent years this question was answered by the systematic search for meteorites in areas favourable for their preservation. From arid areas in parts of the USA and Western Australia the types of meteorites recovered are in roughly the same proportions as the falls in Table 4.1. This shows that the proportions of stones, stony-irons and irons falling to Earth have stayed the same for thousands of years at least. Furthermore, this conclusion is supported by the random meteorite sample conserved for us by the Antarctic ice. It seems, then, that the recovery of irons and stony-irons is enhanced by their unusual appearance and greater resistance to erosion compared with stony types. Boulders and pebbles are commonplace on many land-surfaces; lumps of iron are not. Hence there is a greater likelihood of a finder keeping an iron meteorite than a stony one, the latter being, to the uninformed, unexceptional and just like other rocks.

Each of the three categories, irons, stony-irons and stones, is itself composed of groups of meteorites that differ one from another in chemical composition and mineralogy. This is important, for ultimately we must try to use our

TABLE 4.1
*Meteorite falls and finds*

|  | Stones | Irons | Stony-irons |
|---|---|---|---|
| Falls | 791 | 46 | 11 |
| Finds | 593 | 610 | 67 |
| Total | 1384 | 656 | 78 |

Only verified falls and finds as of December, 1975, are included. Recent finds in Antarctica are not included; these are comprised of the three classes in approximately the same proportions as the falls.

information to deduce how the different types may have been related in space and time. For example, the size and number of the planetary bodies from which meteorites were broken may be tentatively established. Stony meteorites will be dealt with first because, in comparison with irons and stony-irons, their chemistry and mineralogy are more akin to Earth-rocks.

## Stony meteorites

Stony meteorites are divided into two broad classes. The less abundant includes only stones that contain almost no metal or sulphide. Texturally, too, many are not unlike igneous rocks on Earth or on the Moon. This class – the achondrites – has only 85 members (67 falls, 18 finds). The chondrites constitute the more abundant class, numbering 1250 meteorites (685 falls, 565 finds). The names are derived from the Greek, so the 'ch' in each of them is pronounced as 'k'. Almost every chondrite has some metallic iron-nickel and all contain an iron-bearing sulphide. So mineralogically, most chondrites differ from all but the most unusual Earth-rocks. The chondrites are generally composed of small, spherical lumps of silicate set in a groundmass in which most of the metal and sulphide are located (Plate 3a). Such a texture is alien to terrestrial rocks, but, as we shall see, chondrites are important as likely representatives of the material from which planets were built.

The chemical composition and mineralogy of meteorites are intimately bound in a relationship of 'chicken and egg' type: it is usually impossible to determine which came first. So although the modern classification of meteorites is essentially based on chemical composition, in most instances it is equally valid to distinguish the chemical classes of meteorites by the minerals they contain. This alternative approach is used here.

Only eight minerals, either singly or in various mixtures, make up the bulk of all meteorites so far encountered (Table 4.2). About eighty additional species

TABLE 4.2
*The eight most common minerals of meteorites*

OLIVINE    A silicate of magnesium and iron $(Mg,Fe)_2SiO_4$

PYROXENE    A silicate of magnesium and iron, with or without much calcium, and containing more silicon than olivine $(Mg,Fe,Ca)_2Si_2O_6$

PLAGIOCLASE    A silicate of aluminium with varying amounts of sodium and calcium $NaAlSi_3O_8$ to $CaAl_2Si_2O_8$

KAMACITE    Metallic iron-nickel with less than about 7 per cent nickel

TAENITE    Metallic iron-nickel with more than about 25 per cent nickel

TROILITE    Iron sulphide FeS

'SERPENTINE'    Water-bearing silicates of magnesium and iron, for example $(Mg,Fe)_6Si_4O_{10}(OH)_8$

MAGNETITE    An oxide of iron $Fe_3O_4$

of mineral are known to occur in meteorites, but only exceptionally do any of these constitute as much as ten per cent of a meteorite. Most of these minerals are extremely rare, and a number are unknown on Earth. Of the eight common minerals of meteorites, all except the sulphide - troilite - occur on Earth, but the iron-nickel minerals - kamacite and taenite - are exceedingly rare. The remaining five are common terrestrially, especially in oceanic regions.

## The chondrites

The most abundant kind of chondrite contains, by weight, about 45 per cent olivine, 25 per cent pyroxene, 10 per cent plagioclase, 5 per cent troilite and from 2 to 20 per cent iron-nickel metal (kamacite and taenite in varing proportions). The poorest in olivine tend to be richest in metal, and *vice versa*. These meteorites are the ordinary chondrites and account for half of the meteorites known. At least three-quarters by weight of each meteorite is made up of stony minerals, so these are stony meteorites, but it should be remembered that all contain some metal and the sulphide, troilite. This is a more complex mineral assemblage than in rocks occurring at the Earth's surface, most of which have negligible sulphide and no metallic mineral. In structure, too, the chondrites differ from Earth-rocks.

The word 'chondrite' is derived from the Greek 'chondros', meaning a grain. Chondrites are partly composed of rounded grains (chondrules) of silicate that vary in size, say, from that of a pin-head to that of a pea (Plate 3a). Within many chondrules, crystals of olivine and/or pyroxene have forms indicative of very rapid cooling from high temperature (Plate 3b). This was first recognized in the 1870s by the English scientist, Sorby. He noted that the crystals were often of imperfect form, pyroxene being 'feathery' and olivine being skeletal, that is, with a well-formed outline but with an interior filled with glass. The reason is that fast cooling, or quenching, does not allow time for the atoms in the liquid to find their proper sites in the structure of growing crystals, which are consequently full of holes and defects. Experimental work shows that a chondrule-sized drop of molten pyroxene cools rapidly from white heat. Crystallization does not begin until the temperature falls to several hundred degrees below that at which it would begin in a more slowly cooled liquid. But when crystallization does begin it proceeds to completion within seconds, releasing enough heat to cause the drop to glow again momentarily. Many chondrules must therefore have cooled faster than almost all igneous rocks on Earth. Sorby saw that textures within chondrules are like those of slags, which quickly cool when they are discharged as waste products during metal extraction. He likened the formation of chondrites to the fall of a 'fiery rain' which he thought had come from the Sun. This idea is unacceptable today, but after a century, the origin of chondrules and chondrites is still a matter of speculation and debate and is a subject to which we shall return.

Within the ordinary chondrites a threefold division has been established. High-iron – or H-group – chondrites have about 27 per cent total iron and 12 to over 20 per cent iron-nickel metal, by weight. Low-iron – or L-group – chondrites have about 23 per cent total iron and 5 to 10 per cent of iron-nickel metal. Third are the low-iron, low-metal – or LL-group – chondrites with some 20 per cent total iron and 2 per cent iron-nickel metal; these are the least common of the three groups. In addition, the oxygen content *increases* from H- to L- to LL-groups. In the H-group, paucity of oxygen means that the 27 per cent total iron is distributed as perhaps 16 per cent with no combined oxygen and in the metallic state, 3 per cent, again with no oxygen but in the sulphide mineral troilite, and 8 per cent combined with oxygen and located in the silicate minerals. But, at the other extreme in the LL-group, the 20 per cent total iron is distributed between metal, sulphide and silicate in the approximate ratio of $1:3:16$, all in weight per cent. The different levels of availability of oxygen in the ordinary chondrites mean that the silicate minerals of each group normally have closely defined contents of oxidized iron. For example, the weight percentages of oxidized iron in the mineral olivine are roughly 14 per cent in H-group, 17 per cent in L-group and 20 per cent in LL-group chondrites.

The designation of the H-, L- and LL-groups of the ordinary chondrites was erected on the chemical basis just described. Each group may be further subdivided on account of textural and mineralogical criteria.

It was long realized that within each chemical group of the ordinary chondrites the meteorites exhibit a range of textures. In the H-group, for example, the meteorite Tieschitz (fell 1878, in Czechoslovakia) has abundant, well-defined chondrules with dark borders. Space between the chondrules is at a minimum and filled with very fine-grained pale material plus larger fragments of silicate minerals (Plate 3b). Metal and sulphide occur as rounded, chondrule-like blebs and as smaller droplets within chondrules (Plate 3c). In contrast, Kernouve (fell 1869, in France) has a few poorly defined chondrules set in an abundant matrix composed mainly of interlocking silicate crystals, metal and sulphide (Plate 3d). The average grain-size in the Kernouve matrix is much coarser than in the Tieschitz matrix. These two meteorites represent the extremes of texture among the high-iron group chondrites, and between them intermediate types exist.

The relationship is explained as follows. All ordinary chondritic material was originally of the Tieschitz type, with abundant chondrules many of which had solidified during virtually instantaneous cooling from over 1400 °C. Annealing, brought about either by slow cooling or re-heating, allowed crystals to grow. Such crystal growth not only destroyed the quench textures within chondrules, but destroyed also the outlines of chondrules due to grain-growth between chondrules and matrix. This type of process is known as thermal metamorphism: change by the action of heat. Its varying intensity in different parts of the planetary bodies on which the ordinary chondrites formed is thought to have been responsible for the different textural types among the chondrites. In 1967, textural criteria were used by W. R. van Schmus & J. A. Wood, both of the USA, to establish four metamorphic types in the ordinary chondrites.

Van Schmus & Wood recognized four different mineral assemblages indicative of crystallization of ordinary chondrites in different temparature ranges. Unfortunately, their scheme was also applied to another group of meteorites in which occur assemblages of environments of lower temperature than that of Tieschitz. This stone belongs to petrologic type 3. Types 4, 5 and 6 are applied to assemblages of increasingly high temperature, 6 being applicable to Kernouve.

In Tieschitz and other type 3 chondrites, at least some chondrules contain clear, fresh glass (Plate 3b). Pyroxene, because of its rapid crystallization, occurs mainly as crystals of a poorly ordered type extremely rare on Earth. The mineral olivine, too, often occurs in branching (dendritic) forms indicative of fast crystal-growth; its composition is variable, the iron content ranging from zero to 40 per cent by weight. The mineral plagioclase is almost totally absent, being present mainly as a component in the silicate glasses. The chondrules themselves have sharply defined outlines and are distinct from the matrix. Such an assemblage results from rapid cooling of the constituent chondrules from perhaps 1400 °C to 800 °C, or less. Thereafter, temperatures above 700 °C could not have been attained.

Petrologic type 4 is typified by the absence of clear glass and an increase in the ordering in pyroxene crystals. Slower cooling at around 700 °C or re-heating to above this temperature caused sub-microscopic crystals to form in glass, making it appear turbid under the microscope. Olivine crystals have a uniform composition within each meteorite, but pyroxenes are still somewhat variable both structurally and chemically. Chondrules are slightly less well defined than in type 3.

In type 5, chondrules are considerably less pronounced than in the lower types, and not even turbid glass is present. Plagioclase occurs as small, cloudy crystals visible under the microscope, and pyroxenes are structurally uniform and almost entirely of the well-ordered type (Plate 3e). To attain this texture, the meteorite must either have cooled very slowly or else it was re-heated to perhaps 850 °C after it had formed cold.

Chondrules are well integrated wih the crystalline matrix in type 6 chondrites. Plagioclase is present in easily identifiable, clear crystals (Plate 3d), and there is greater uniformity of texture than in the lower types. Petrologic type 6 is consistent with extremely slow cooling or re-heating to 950 °C.

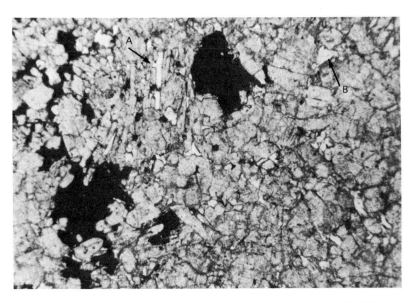

FIG 4.1  Photomicrograph of the Shaw, Colorado, L7 chondrite. Width of field 2·4 mm. When a rock or stony meteorite is cut thinly enough, most of the minerals become transparent and the thin-section, usually 0·03 mm thick, may be studied under the microscope. Note the granular, wholly crystalline texture. The only suggestion of a relic chondrule structure is seen at A. This was a type comprising bars of the mineral, olivine (grey), with areas of plagioclase (white). In chondrites of petrologic types 3 or 4, such areas are either glassy or turbid and finely crystalline, their outline being determined by the enclosing olivine. In the figure, plagioclase is also indentifiable at B.

Finally, one L-group chondrite and fragments occurring in some L- and LL-chondrites have no chondrules at all. Here, the mineral grains form a completely interlocking, granular mosaic, which is often referred to as type 7 (FIG 4.1). Such a texture was probably attained by prolonged heating at 1200 °C.

At this temperature, metal and sulphide melted and began to drain away from the stony material.

Most ordinary chondrites have been annealed or re-heated, petrologic types 5 and 6 being abundantly populated. In contrast, of 523 H-group chondrites known in 1975, only 14 were positively identified as belonging to type 3. The controls governing the distribution of the ordinary chondrites among the different petrologic types touch on some of the most important unresolved questions on solar system history. In particular, how were the parent-bodies of chondrites kept hot or re-heated? Do petrologic type 3 chondrites come from near the surface or near the centres of their parent-bodies? (see Chapter 6, p. 111) Questions such as these form a recurring theme of this book.

Readers will no doubt be curious about the identity of petrologic types 2 and 1. These types apply to even lower temperature mineral assemblages than occur among the ordinary chondrites. Both types are typified by the presence of water-bearing minerals which, in meteorites, are confined to the group known as carbonaceous chondrites. These are important for our search because not only do they possess water, which is essential to life on Earth, but they also contain complex chemical compounds of carbon, hydrogen, nitrogen and oxygen. Such chemicals, if not themselves the forerunners of living organisms, may have been food for living things.

Because a number of carbonaceous chondrites do not contain water-bearing minerals, their presence is not diagnostic of the group. What the meteorites have in common are, relative to silicon, contents of magnesium, aluminium and calcium higher than in the ordinary chondrites. For every atom of silicon, a carbonaceous chondrite has 1·05, or more, atoms of magnesium. In addition, carbonaceous chondrites are rich in oxygen and, apart from one exception, contain either no iron-nickel metal or almost none. These meteorites are dark-looking rocks which appear not unlike some of those on Earth. *All* carbonaceous chondrites are mixtures of minerals formed at different temperatures.

The carbonaceous chondrites are divided into four groups, the most abundant being designated CM2 with about 18 members, all but two of which are observed falls. 'C' stands for 'carbonaceous', 'M' is for 'Mighei', a meteorite which fell in the Ukraine in 1889 and is now represented in some of the world's major museums, and '2' is for the petrologic type. CM2 stones have a matrix of 'serpentine', a group of water-bearing silicate minerals, plus smaller amounts of the iron oxide, magnetite, epsomite (magnesium sulphate, or epsom salts) and other low temperature minerals. Set in the matrix are chondrules, isolated olivine grains, some perfectly formed (FIG 4.2), and rare crystalline aggregates of high temperature minerals (FIG 4.3).

Another important group of carbonaceous chondrites are the CIs, only 5 of which are known. Again, 'C' is for carbonaceous and 'I' is for Ivuna, a Tanzanian fall representative of the group, which is typified by the absence of chondrules and the abundance of low temperature, largely water-bearing minerals (Plate 4a). These meteorites contain about 20 per cent water compared to 10 per cent

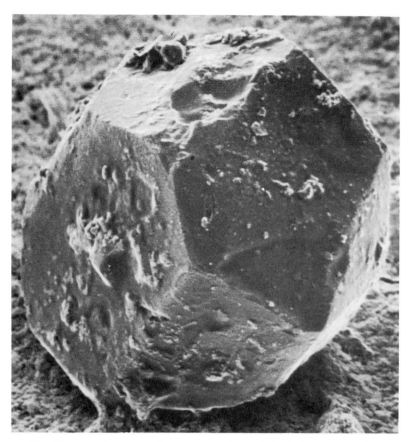

FIG 4.2   Electron micrograph of an almost perfectly formed olivine crystal from the Murchison, CM2, carbonaceous chondrite. Width of field 0·4 mm. The scanning electron microscope was used because it gives better resolution and better depth of field than the optical microscope. Each face is well developed, which suggests that the crystal grew by the addition of atoms condensed from a gas while the crystal was in free flight. Had the crystal been attached to a surface, growth of the attached faces would have been inhibited, as with the closest terrestrial analogues.
(Courtesy Drs E. J. Olsen and L. Grossman, Field Museum of Natural History and University of Chicago).

in CM2s. Although it may at first seem illogical to class these chondrule-free stones as chondrites, this is not so, because the CIs have chemical affinities with the CM2s, which do have sparse chondrules. Indeed, the minerals of the CM2 matrix constitute the bulk of the CIs which are, you will guess, composed of 'serpentine', magnetite and epsomite. Epsomite dissolves readily in water, which

FIG 4.3    The Murchison, CM2, carbonaceous chondrite. Width of field 4 cm. Note the light coloured chondrules (A) and angular high-temperature inclusions (B) set in a dark matrix. The matrix is composed mainly of water-bearing, 'serpentine'-like minerals, but contains 'organic' compounds such as amino acids and complex, tarry substances, plus magnetite, epsomite, etc. (see text and also see Figure 7.7).

may explain its occurrence in veins in CI chondrites – it appears to have been deposited by liquid water. In these meteorites, metal is absent, the only high temperature mineral being olivine which is present as tiny, isolated, well-formed crystals.

In planetary science, the significance of the CIs bears no relationship to their small number, for these five stones have a chemical composition that matches closely the 'condensable' portion of our Sun. They form the basis of our estimate of 'cosmic' composition. What this means is that our theories of the origin and evolution of the solar system into Sun, planets, comets and asteroids mostly begin by assuming that the non-gaseous portion of the whole has an average composition like these five meteorites. A further point of interest is the suggestion that CIs and CM2s are related to comets. The argument is as follows: the light emitted by meteors and their friability during atmospheric flight indicate that the objects responsible have compositions like those of the two meteorite

FIG 4.4   The Vigarano, CV3, carbonaceous chondrite, width 15 cm. On this polished surface are clearly visible large, sub-circular chondrules and white, high-temperature inclusions with irregular outlines.

groups. Radar and photographic observations indicate that the orbits of meteors are cometary. So CIs and CM2s are probably cometary also. If this is the case, it helps to explain the rarity of these meteorites, because a large proportion of cometary debris intercepts the Earth's orbit at much higher velocity than material from the asteroids; and high velocity encounter with our atmosphere is totally destructive of friable objects.

As well as embracing meteorites with no chondrules, the carbonaceous chondrites include over a dozen meteorites with little carbon. These are linked to the CM2s because of the high magnesium to silicon ratio and high oxygen content. There are two types which were at first distinguished largely on a textural basis; later, chemical differences were shown to exist. One type, the CV3s, is typified by the presence of large chondrules which often contain sulphide and metal, and the ratio of chondrules to matrix is low. 'V' is for the type meteorite, Vigarano, and '3' for the petrologic type. There are also four CV2s, with water-bearing matrix minerals. In addition to chondrules, CV meteorites also have irregular, whitish inclusions set in an olivine-rich matrix (FIG 4.4). The

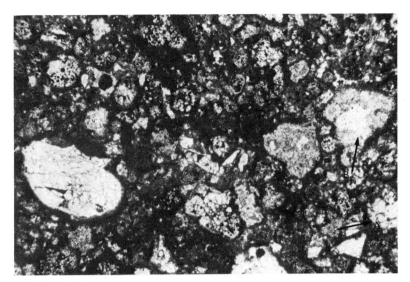

FIG 4.5   The Ornans, CO3, carbonaceous chondrite in thin-section. Width of field 2·4 mm. Tiny chondrules (an olivine-rich example appears at A) and angular, high-temperature inclusions (B) are packed together with a little, fine-grained matrix.

inclusions are composed of calcium- aluminium- and titanium-rich minerals originally considered to have condensed at high temperature from a solar nebula during the formation of the solar system. In contrast, a rival theory suggests that the material in the white inclusions may be from the outer layers of an exploding star which triggered solar system formation. If this is correct, these are probably the only objects accessible to man that have come from outside the solar system. Whatever the correct explanation, it is possible that these centimetre-sized inclusions are the oldest pieces of rock at our disposal. But, more recently, various origins have been proposed, depending on their compositions and textures. We return to them in detail in Chapter 7.

The other type of carbon-poor carbonaceous chondrite is called 'CO3'; 'O' for the Ornans stone, '3' for the petrologic type. Only seven are known. They are characterized by a great abundance of densely packed, small chondrules less than 0·2 millimetres in diameter (FIG 4.5). Some chondrules have mineral assemblages like those of the white inclusions in the CV3 meteorites, but in the Ornans type it is evident that the minerals have been reworked. One CO4 meteorite is known; it fell at Karoonda, South Australia, in 1930.

Until 1972, no highly metamorphosed carbonaceous chondrite was known, but in that year a stone weighing only 169 grams was found on the Nullarbor Plain in Western Australia. Subsequent investigation revealed that although it has the bulk composition of a carbonaceous chondrite, chondrules are poorly defined and the olivine is of uniform composition. In short, the stone has the

FIG 4.6   Polished surface of the Khairpur, E6, enstatite chondrite. Width 4 cm. Note the abundance of shiny iron-nickel metal, the distribution of which is somewhat uneven. A few metal-free, oval grey areas are indicative of relic chondrules.

characteristics of petrologic type 5 or 6. The discovery of this small stone, less than the size of a man's fist, has meant that we must now accept that one parent-body of a carbonaceous chondrite was either strongly re-heated or slowly cooled shortly after its formation. This appears to destroy the theory of a cometary origin for at least one carbonaceous chondrite – unless it tells us something about comets?

We now move on to consider the enstatite chondrites, the group of meteorites the origin of which is the most difficult to explain. Only twenty are known, eleven being observed falls. They contain so little oxygen that all of the iron occurs as metal or as troilite. A further diagnostic feature is the low magnesium to silicon ratio. In the mineral olivine, there are two atoms of magnesium or iron to each silicon atom. But in enstatite chondrites no iron occurs with oxygen in silicate minerals, and for every atom of silicon there is, on average, only about three-quarters of a magnesium atom. So, unlike other chondritic groups, the mineral assemblage of these meteorites is dominated by the magnesium silicate, enstatite, which gives the group its name. This mineral is a pure magnesium pyroxene, with an atomic magnesium to silicon ratio of 1:1. Because of the extreme paucity in oxygen, elements such as calcium, titanium and chromium occur in combination with sulphur. Even some potassium and sodium are present in sulphides, an occurrence almost totally unknown on the oxygen-rich Earth. The enstatite chondrites are the home of a weird, oxygen-depleted group of very rare minerals. But the more abundant minerals are common enough in meteorites: perhaps 65 per cent enstatite, 15 per cent kamacite, 10 per cent troilite and 5 per cent plagioclase (FIG 4.6). In this group, six stones are chondrule-rich and placed in type 4, ten are chondrule-poor and assigned to type 6, two are of type 5, and in addition, the unique meteorite, Happy Canyon, is type 7. (One enstatite chondrite has not yet been properly classified.)

As with the carbonaceous chondrites, the importance of the enstatite chondrites is greater than their small number would suggest. For example, relative to chondrites, the Earth appears to be impoverished in sodium and potassium. It is often argued, by analogy with the enstatite chondrites, that on Earth these elements are concentrated in the outer core, where they occur in association with sulphur. Furthermore, it seems that the outer layers of the Earth are depleted in silicon also. Once again, by analogy with the presence of metallic silicon dissolved in the metal of enstatite chondrites, it has been suggested that part of the Earth's silicon is locked in the metallic core. However, all this is largely speculation.

The main problem with the enstatite chondrites is our lack of understanding of the conditions in which they could have formed. We know that on Earth and Moon the accessible rocks contain more oxygen than the enstatite chondrites. The atmosphere and surface of Mars contain water and carbon dioxide, both indicative of plentiful oxygen. Venus, too, has an atmosphere that looks to be rich in oxygen combined in carbon dioxide or in sulphuric acid. Beyond Mars, conditions in the early solar system should have been colder than in the vicinity of Earth, and hence the objects forming further out should have been more highly oxidized than Earth. By elimination, we are forced to accept that the enstatite chondrites may have formed closer to the Sun than Venus. You will recall that Mercury is a dense planet for its small size, indicating that it is rich in metal and poor in oxygen, characteristics shared with the enstatite chondrites. But this only leads us to a further problem. Enstatite chondrites still fall to Earth now and then. If they formed in the vicinity of Mercury, at least one enstatite chondrite body must have been ejected from an orbit in the inner solar system into a new orbit further out. This new orbit must then have been stabilized, presumably in the asteroids, to allow debris produced in sporadic collisions to be sent into Earth-crossing orbits. Such a history is dynamically improbable.

The classification of chondritic meteorites is summarized as Table 4.3.

We have just examined the chondrites. Although there are mineralogical, chemical and textural differences between the various kinds, they have one thing in common. They were never melted. The chondrites are, therefore, probably representative of the bulk material accumulated on different bodies in various parts of the solar system. Not so, the remaining meteorites, all of which are formed of minerals selectively extracted from some primary materials. Such a

TABLE 4.3
*Summary of chondrite classification*

|  | Type 1 | Type 2 | Type 3 |
|---|---|---|---|
| **Carbonaceous** Mg/Si atomic greater than 1·05 | Chondrules absent | Chondrules sparse | Chondrules abundant |
|  | 'Serpentine' and other water-rich minerals are abundant | | Matrix is fine— |
|  | Olivine very rare | Composition of olivine variable | |
|  | CI | CM2 small chondr. | CO3 small chondr. |
|  |  | CV2 large chondr. | CV3 large chondr. |
| **Ordinary** Mg/Si atomic =0·95 LL Total Fe 20% Metal 2% L Total Fe 23% Metal 8% H Total Fe 27% Metal 20% |  |  | Chondrules sharply defined |
|  |  |  | Matrix opaque, fine-grained |
|  |  |  | Glass clear |
|  |  |  | Olivine of variable comp. |
|  |  |  | Pyroxene mainly disordered |
|  |  |  | Plagioclase is |
| **Enstatite** Mg/Si atomic less than 0·85 |  |  |  |

process, in which one portion, or fraction, of material is separated from another, is called 'fractionation'. To explain the chemical differences (in total iron and in oxygen content) between the different groups of chondrites we assume some kind of separation process between gas and dust. This took place at about the time of formation of the planets. Dust accumulated on to growing planets; most of the gas was subsequently blown out of the inner solar system by the Sun's radiation; the remainder of the dust either spiralled in towards the Sun or, if fine enough, was expelled with the gas. The dust and condensed gases left behind as planetary bodies vary in composition due to the action of fractionation processes. For example, magnetized metal grains tend to be attracted and so stick together more readily than stony grains. Metal would, therefore, have tended to settle early on to growing bodies. Early-formed bodies would be metal-rich, late-formed ones, metal-poor. Some of the bodies, we believe, had chondritic compositions. Strong heating of a chondritic body causing melting would have allowed metal to sink to form a core, displacing molten rock which would have tended to move upwards. Most meteorites other than chondrites

| Type 4 | Type 5 | Type 6 | Type 7 |
|---|---|---|---|
| and distinct | Chondrules indistinct | | |
| olivine-rich grained | Matrix coarse, granular | | |
| Olivine has uniform composition | | | |
| CO4 small chondr. | | | |
| CV4 large chondr. | CV5 large chondr. | | |
| Chondrules distinct | Chondrules not distinct | Chondrules sparse, indistinct | Chondrules absent |
| Matrix opaque, less fine-grained | Matrix granular | Matrix granular, coarse-grained | Coarse-grained texture throughout |
| Glass turbid | Glass completely absent | | |
| Olivine has uniform composition throughout | | | |
| Pyroxene mainly ordered | Pyroxene is ordered (orthorhombic) throughout | | |
| totally absent | Plagioclase in turbid crystals | Plagioclase clear, in well defined crystals | |
| | Olivine is almost completely absent. Criteria for petrologic grades 4–7 of the ordinary chondrites apply | | |

are thought to have formed in such a melting process. The chondrites themselves may come from bodies small enough to have escaped melting, or, alternatively they may be from near-surface areas that never melted.

## The achondrites

Stony meteorites apart from chondrites are known as achondrites, which means 'without chondrules'. The achondrites generally are igneous rocks, although one group was formed by the mixing of mineral and rock fragments as in a lunar 'soil'. We shall begin by discussing the eucrites, which are meteoritic basalts, but before we do so it is worth considering some of the implications of melting in small bodies.

The chondrites were produced by the accumulation of chondrules and mineral fragments on planetary surfaces. As just stated, most achondrites are igneous rocks and the product of melting on small, planetary bodies. We do not

FIG 4.7   The Sioux County eucrite. Width 7 cm. An igneous meteorite composed of basaltic fragments and broken mineral grains. A coarse-grained fragment (arrowed) is an intergrowth of crystals of dark pyroxene and white plagioclase. Note the dark, shiny fusion-crust (top left).

know what caused these bodies to melt. Impact is the most convenient explanation, because other sources of energy require more uncertain conditions, but impact is a most inefficient means of melting rocks, most of which are merely pulverized. The available evidence indicates that igneous achondrites, with few exceptions, all crystallized at about the same time some 4500 million years ago. Thus, if impact were the energy source, it caused melting at one time only and on several bodies of differing composition. (The exceptions include the Nakhla stone, an igneous rock formed about 1200 million years ago.) The lunar crust, too, was formed by partial melting at roughly the same time. This contemporaneity of heating is unlikely to have been produced by a random process such as impact. We are then forced to accept that a process other than impact was responsible for the melting by which most of the achondrites were produced, which is something of an embarrassment, for scientists are undecided on how the melting could have been brought about.

That melting *did* occur is attested by the eucrites. This group comprises thirty meteorites that closely resemble lunar lavas and are not unlike terrestrial

FIG 4.8    The plagioclase cumulate eucrite, Serra de Magé, in thin-section. Width of field 2·6 mm. An elongate, well-formed plagioclase crystal is surrounded by granular crystals of the same mineral, except for three pyroxene crystals (Px) and the opaque sulphide mineral, troilite (far right, black). Crossed polars (for explanation, see caption to Plate 1b).

basalts. Eucrites are composed of approximately equal proportions of plagioclase and a calcium-bearing pyroxene (FIG 4.7). Minute amounts of iron-nickel metal and troilite are always present, this being in common with lunar lavas. But eucrites do not come from the Moon, for they have suffered radiation damage consistent with a longer exposure to cosmic rays than would have been incurred on the relatively short journey from Moon to Earth. The major difference between eucrites and terrestrial rocks is that the former have very much lower sodium and potassium contents.

A few eucrites are enriched in either plagioclase or pyroxene. Accumulation of crystals of one or other mineral by settling from eucritic liquid almost certainly is the process responsible. Imagine a body of molten rock. Most of the cooling occurs through the top, causing crystals to form there. These gradually sink to the bottom, where they accumulate. If, for example, only plagioclase is crystallizing from the melt, a layer of plagioclase-rich rock forms. Such a rock is called a plagioclase cumulate, and is exemplified by the Serra de Magé eucrite (FIG 4.8).

Five classes of achondrite are cumulates of different sorts. The hypersthene achondrites, ten in number, are composed of the mineral, hypersthene, which is a type of pyroxene with little calcium and about 12 per cent of iron by weight (Plate 4b). It was originally thought that these meteorites represent early-formed hypersthene crystals from eucritic liquids, but melting experiments carried out

FIG 4.9    Kapoeta, Sudan, howardite. Width of field 6·5 cm. The meteorite is essentially a 'soil' from a planetary surface, and contains fragments of eucrite (A) and other components. Part of the meteorite is dark because of solar irradiation, the remainder being light and unirradiated. The junction between 'light' and 'dark' is visible (B), the former being down and to the left.

in Edinburgh have shown that when a eucrite is melted and allowed to cool slowly, olivine, not pyroxene, is the mineral to crystallize first. The hypersthene achondrites must therefore have formed by the accumulation of crystals from a liquid of composition different from that of any known eucrite.

Another group of achondrites – the howardites – serves as a link between the eucrites and hypersthene achondrites. The howardites formed on a planetary surface as 'soils' rich in fragments of eucrite and hypersthene achondrite, plus a minor chondritic component (FIG 4.9). Clearly, material from the eucrite and hypersthene achondrite parent-body, or bodies, was simultaneously eroded, probably by impact. The eroded material was mixed with incoming chondritic fragments and deposited to form a 'soil'. All of this possibly took place on a single planetary body, the process of formation being analogous to the formation of 'soils' on the Moon. Like the Moon, the parent-body of the howardites probably did not have an atmosphere, for some crystals in the howardites have

FIG 4.10    Thin-section, Hajmah (a), Oman, ureilite. Width of field 2·4 mm. A cumulate composed of olivine (Ol) and pyroxene (Px) crystals with dark, carbon-rich (C), graphite- and diamond-bearing material in between.

been damaged by the particles from the solar wind. The damage to crystals usually occurs on all sides indicating that the crystals were turned around as they were subjected to solar radiation. This discovery was made in 1969 before the first Moon-rocks had been collected. At that time it was suggested that the grains in the meteorite had been irradiated during free flight in space, before they landed on the howardite parent-body. However, Apollo XI 'soils' were found to contain crystals with identical damage, and in the lunar case the crystals were certainly not in free flight. Over long periods of time, crystals on the Moon's surface are turned over by a 'gardening' process due to impact. It is now thought that the howardites formed in the same way. If this is so, we can safely say that the howardites, eucrites and hypersthene achondrites formed in the inner part of the solar system. This is because the solar wind becomes progressively weaker further from the Sun. Jupiter, for example, is just over five times more distant from the Sun than Earth. At Jupiter, the strength of the solar wind is only about one twenty-seventh of its strength at Earth; at Saturn, it is less than one ninetieth. So, if the howardites formed in the same way as lunar 'soils', with grains spending similar times on the surface, the parent-body of the howardites certainly formed nearer to the Sun than the orbit of Saturn. Spectroscopic studies indicate that one parent-body may be the asteroid, Vesta.

Another interesting group of achondrite cumulates are the ureilites. For many years only three were known, then, in the mid 1960s, two different ones were found within about 30 kilometres of each other in Western Australia. The total now stands at eight. Ureilites are cumulates of mainly olivine crystals, with minor amounts of a calcium-bearing pyroxene (FIG 4.10). What is remarkable is the presence of carbon, some of it as diamond. Now diamond forms only at high temperatures and pressures and in the ureilites these conditions were achieved by shock. The small, planetary home of the olivine-pyroxene cumulates was involved in a collision in space. Temperature and pressure rose momentarily, and diamond was formed from some of the graphite, the low pressure form of carbon. At high temperature, some of the oxidized iron in the outer parts of olivine crystals was reduced to metal. The carbon seems to have acted like coke in a blast furnace, so some of the original carbon must have been converted to carbon monoxide or carbon dioxide, and lost. The ureilites, then, were a sort of diamond-producing blast furnace in space, but, unfortunately, the diamonds are microscopic.

We have looked briefly at four of the eight achondrite groups. The fifth, the enstatite achondrites, is worth mentioning as it is not unlike metal and sulphide-poor, coarse-grained enstatite chondrite material. But the relationship between chondrite and achondrite is still obscure. The remaining three groups contain only six meteorites between them, so in terms of numbers are relatively unimportant. However, one of the six, the pyroxene-olivine cumulate, Nakhla, has the youngest crystallization age of any meteorite – 1200 million years (Plate 4c). It has several features in common with terrestrial igneous rocks, such as the presence of oxides of iron and titanium and a small amount of water. It has been suggested that Nakhla may be the product of volcanism on a large, active planet such as Mars. However, the meteorite is unshocked, and it is improbable that such a crystalline rock could have been accelerated to the escape velocity of Mars, 5 km per second, without destroying the texture.

### Iron meteorites

Like Nakhla, many iron meteorites appear to have formed as cumulates. But in the case of the irons, the crystals were of iron-nickel metal and the liquids most probably were the molten metal cores of small planetary bodies. However, there is evidence that a minority of iron meteorites are composed of material that was never completely molten.

Iron meteorites are dominantly composed of the two iron-nickel alloys, kamacite and taenite. About fifty irons have between 5 and 6·5 per cent of nickel by weight. These contain only the low-nickel mineral, kamacite, with minor amounts of the sulphide, troilite. Because they are formed of cubic crystals of kamacite they are called 'hexahedrites', after the Greek 'hexahedron' for a cube, which has six faces. At the other extreme, a small number of irons with more than 13 per cent of nickel, together with a few others, are composed of the high-nickel alloy, taenite, to the almost complete exclusion of low-nickel ka-

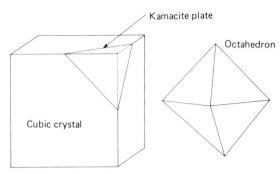

FIG 4.11   Relationship between the cube and the octahedron. In most iron meteorites kamacite occurs as four sets of plates parallel to the four octahedral planes of the parent taenite cube.

macite. They are structureless and are known as 'ataxites', meaning 'without order', again from Greek. Irons that have nickel contents beteen 6·5 and 13 per cent are generally composed of plates of kamacite set in a taenite matrix. This is true of the majority of irons, which are called 'octahedrites' for the following reason:

Solid iron-nickel metal at over 800 °C has the crystal structure of the mineral taenite. When an alloy with between 6·5 and 13 per cent of nickel cools to lower temperatures, the taenite structure becomes unstable. To maintain stability, plates of low-nickel kamacite form within the original cubic crystal of taenite. The remaining taenite becomes smaller in quantity but richer in nickel as cooling continues. Kamacite is present in four sets of parallel plates that cut across the corners of the original taenite cube. The four sets are parallel to the faces of the eight-faced solid known as an octahedron (FIG 4.11).

At the beginning of the nineteenth century it was found that many iron meteorites, when polished and treated with acid, exhibit a criss-cross pattern (FIG 4.12). This is known as the Widmanstätten pattern, after the name of the Austrian count who was one of its discoverers. The pattern is the result of preferential etching of kamacite plates in the octahedrite structure.

Until recently, iron meteorites were classified on a purely structural basis as hexahedrites, octahedrites or ataxites. The octahedrites were further subdivided depending on the thickness of the kamacite plates – compare Figures 4.12 and 4.13. *In general*, nickel content increases from hexahedrites, to octahedrites with coarse kamacite plates, to octahedrites with fine kamacite plates, to ataxites, which are richest in nickel. Thus the old structural classification had *some* chemical basis. However, analytical work begun in the 1950s and reaching a high level of precision some tens years later, has led to a purely chemical classification for the irons. Professor Wasson of the University of California at Los Angeles was mainly responsible for showing that most iron meteorites

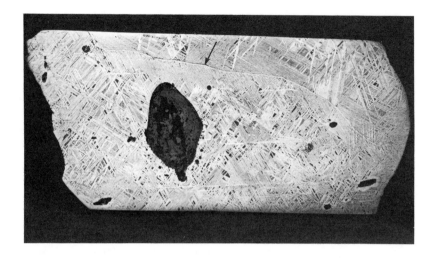

Top, FIG 4.12 Polished and etched face, Gibeon, Group IVA, iron. The line running from top left to lower right (arrowed) is the boundary between two parent taenite crystals. In each parent taenite crystal kamacite plates may be seen as light coloured lines running in four different directions. The large, oval dark area is composed of the iron sulphide, troilite. Structurally, Gibeon is a fine octahedrite. Compare its thin kamacite lamellae with those in the coarse octahedrite, lower illustration. Width 14 cm.

belong to one or other of twelve chemical groups (FIG 4.14). Unfortunately, in addition to nickel, this scheme is based on the abundances of the metals gallium, germanium and iridium which are present in small amounts and so are difficult to determine. Because of this many iron meteorites can be classified by only a few specialized research laboratories. But fortunately Wasson worked in collaboration with a Danish metallurgist, Dr Buchwald. Detailed study of the structure of the iron-nickel alloys and of the mineralogy enabled Buchwald to relate chemistry to microscopic observation. So now a newly discovered iron may be quickly and cheaply assigned to one of the twelve chemical groups, or else recognized as something exceptional.

But what of the twelve groups of irons? At the least, they indicate that iron meteorites are probably derived from not less than twelve parent-bodies. This is because the chemical and mineralogical differences between the groups are so large that it is improbable that any two could have formed due to variation in temperature and pressure in a single planetary body. Let us now examine in detail some of the evidence for this.

The first attempt at setting up a chemical classification of iron meteorites resulted in the establishment of four groups – Group I with highest gallium and germanium, ranging downwards to Group IV in which these elements were immeasurably low. Subsequent work indicated that Group I should be split into two groups, named IA and IB; still later and more comprehensive study showed that these 'groups' are intergradational and in fact belong to a large, single group now called IAB. And so on for the other groups. We are now in the somewhat confusing position that each pair of sub-groups IA and IB, IIA and IIB, IIIA and IIIB are genetically related and known as IAB, IIAB, IIIAB. In contrast, groups IVA and IVB are very different and totally unrelated. But there are only twelve groups, so their different identities are not difficult to remember.

To understand the significance of the groups we shall discuss groups IAB and IIAB. Group IAB contains 19 per cent of all properly studied iron meteorites, and Group IIAB some 11 per cent. From Figure 4.14 it is seen that IAB irons have nickel contents between 6.4 and 25 per cent. As nickel increases gallium and germanium decrease, from 100 to 11 and 520 to 25 parts per million, respectively. (1 part per million (ppm) is one gram per tonne.) Structurally, IAB irons range from coarsest octahedrites, that is, irons composed mainly of kamacite, to a single ataxite, San Cristobal, which is dominantly taenite. Many members of the group contain inclusions of stony minerals. When we plot nickel content against gold we see (FIG 4.14) that gold remains at rather less than 2 ppm over the whole wide range of nickel. Group IIAB is a much more coherent

FIG 4.13   Polished and etched face, Cañon Diablo, Group IAB, iron; a fragment of the body that produced Meteor Crater. Scale as in FIG 4.12, the field is 14 cm across. The top part has not been etched. Note the thick kamacite plates, one set of which is almost parallel to the face and so members appear broad and irregular. A troilite nodule appears on the lower left, surrounded by carbide and phosphide minerals typical of the chemical group (IAB).

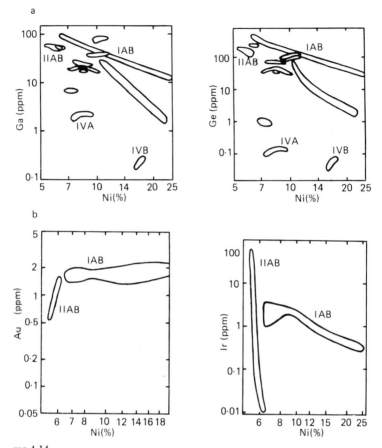

FIG 4.14

a) Nickel-germanium and nickel-gallium plots of iron meteorites. These are the basis of the chemical classification of the irons, almost 90 per cent of which plot within one of the groups; the remainder are termed 'anomalous'.

b) Nickel-gold and nickel-iridium plots of irons of groups IAB and IIAB. In the irons of Group IIAB, gold increases in abundance with that of nickel; in contrast, although Group IAB irons have a wide range of nickel contents, their gold content is essentially constant. Iridium is a noble metal that is chemically similar to platinum. In IIAB irons iridium (and platinum) greatly decreases as nickel increases, whereas there is less variation in the iridium content of IAB irons. For interpretation, see the text.

From Scott and Wasson, courtesy American Geographical Union.

group than IAB. In IIAB, nickel has the limited range from 5·3 to 6·4 per cent, which means that these irons are either hexahedrites (all kamacite) or coarsest octahedrites (with a little taenite). Gallium and germanium, too, have limited ranges (FIG 4.14). However, gold content extends from 0·5 to 1·5 ppm, and increases with increasing nickel abundance (FIG 4.14).

Differences in chemical composition between the two groups of meteorites are generally attributed to their having had different histories. Group IAB probably was never fully molten; IIAB members are parts of the once molten core of a small planetary body. Our reasoning is that slow cooling of iron-nickel liquid will first produce an iron-rich solid, leaving the residual liquid enriched in nickel. Gold follows nickel and tends to stay in the liquid. Group IIAB irons which are poor in nickel are also poor in gold and probably represent early formed solid from a slowly cooling and crystallizing core. In contrast to gold, the metal, platinum, tends to follow iron. Group IIAB irons poor in nickel are highly enriched in platinum, adding weight to our argument. As the core continued to solidify, successive solids become richer in nickel and gold, but depleted in platinum. With Group IAB the very large range in nickel coupled with almost no variation in gold or platinum argues against a fractional solidification mechanism. Besides, had IAB metal been molten, any inclusions of silicate would have floated out, yet many IAB irons of different nickel contents contain small amounts of silicate.

## Stony-irons

In addition to irons and stones, some eighty meteorites are classed as stony-irons. These constitute two groups. The larger is the pallasites, the members of which are composed of clusters of olivine crystals set in metal with the octahedrite structure (Plate 4d). The most plausible explanation for their chemistry and texture seems to be that liquid metal was forced upwards into a cooling and, therefore, contracting and cracking mass of olivine crystals. Arguments, like that above for IIAB irons, suggest that the molten metal may have been the residual liquid after the solidification of eighty per cent of a planetary core. From this core were derived the Group IIIAB irons. Unlike the pallasites, the other major stony-iron group, the mesosiderites, are a mixture of materials from several, possibly unrelated, sources. They contain angular chunks of rock composed of plagioclase and calcium-bearing pyroxene. These are probably fragments of eucrite. Rounded masses of olivine also occur, and the metal component is present largely as round slugs. However, some of the metal occurs as veins, indicative of high pressure, shock-mobilization due to impact (FIG 4.15).

## Summary

We have now discussed the main groups of meteorites. The abundant, ordinary

FIG 4.15    Polished surface of the Crab Orchard mesosiderite. Width 26 cm. Round slugs, angular fragments and veins of shiny metal are unevenly distributed. Silicates are dark and occur as rounded to angular fragments and as finer intergrowths with metal. Contrast the even texture of a pallasite (Plate 4d). Mixing of metal and silicate to form the mesosiderites must have been extremely violent.

chondrites are composed of silicate and minor metal. Although never molten, most experienced either slow cooling from an initial high temperature or a re-heating episode. The carbonaceous chondrites are metal-poor, high and low temperature mineral assemblages many of which contain water and carbon compounds. Since their formation, all but one have stayed pretty cold. The enstatite chondrites contain less oxygen than the other groups and are metal-rich and magnesium-poor. Most were either slowly cooled or re-heated.

The achondrite groups represent the products of melting. The most familiar are the eucrites which are similar to lunar basalts. The howardites apparently formed on a planetary surface or surfaces by a process akin to that which forms lunar 'soils'. With few exceptions, achondrites are fractured, this being a testimony to their violent histories near planetary surfaces.

Like many achondrites, most iron meteorites are the product of melting, probably in the cores of small planets. Chemically, irons are divisible into twelve groups, this grouping being thought to have genetic significance. Differences in nickel content led to the formation of different structures in the metal, for example, low-nickel irons are dominantly of kamacite with hexahedrite structure.

Lastly are the stony-irons. Pallasites are the product of gentle mixing of

**TABLE 4.4**
*Meteorite classification*

| Stony | | | | | Stony-irons | Irons | |
|---|---|---|---|---|---|---|---|
| Achondrites | Chondrites | | | | | (12 chemical groups) | |
| | Carbonaceous | Ordinary | Enstatite | | | | |
| Eucrites | CI (type 1) | H (types 3–6) | 6 chondrule-rich (type 4) | | Mesosiderites | I AB | IC |
| Howardites | CM (type 2) | L (types 3–7) | 2 intermediate (type 5) | | Pallasites | II AB | IIC |
| Hypersthene achondrites | CV (types 2–5) | LL (types 3–7) | 10 chondrule-poor (type 6) | | | II D | II E |
| Urelites | CO (types 3–4) | | 1 highly crystallized (type 7) | | | IIIAB | IIICD |
| Enstatite achondrites | | | | | | IIIE | IIIF |
| Nakhlites | | | | | | IVA | IVB |

Generally decreasing Ga and Ge →

molten metal with neighbouring olivine crystals. Mesosiderites are mixtures also, but formed violently from materials from different sources.

Among meteorites, melting, heating or slow cooling were the rule rather than the exception. Further evidence indicates that the parent-bodies of meteorites must have had a source of heat different in character from any existing in the solar system now, but all meteorites originated in the solar system.

# 5
# Chemical similarities and differences

We discussed in Chapter 3 the Moon and inner planets, and in Chapter 4 the different types of meteorite were presented. All of these have something in common – they share the same Sun. In this chapter we compare and contrast meteorites, inner planets, Moon and Sun in terms of their chemistry and try to suggest processes by which chemical differences could have been achieved.

The Sun weighs almost one thousand times more than all of the planets and their satellites put together. Thus, if we can determine the chemical composition of the Sun we will have obtained a good approximation of the chemical composition of the whole solar system. Unfortunately, in practice this is very difficult to accomplish. For the Sun, like other stars, is a nuclear reactor. But unlike man-made reactors which produce energy by fission, at the high temperatures and pressures obtaining near the centre of the Sun, atoms of hydrogen fuse together to form helium. The helium so produced by fusion weighs very slightly less than the total weight of the hydrogen atoms from which it was made. This difference in weight, or mass annihilated, represents a huge release of energy – enough to drive the Sun. Here lies the root of our problem. The Sun's energy is generated near the centre, but it has to travel through the outer layers of the Sun before it is emitted as radiation. It is because we know little of the internal structure of the Sun or of the mechanisms of energy transfer that we have difficulty in measuring the composition.

Atoms of each chemical element absorb or emit energy at particular levels. When energy in the form of light is applied to cool atoms, the atoms absorb light at particular wavelengths. Conversely, energetic, hot atoms absorb light at particular wavelengths and emit energy as light at other wavelengths. Throughout the Sun, absorption and emission allow the transfer of energy upwards, but other mechanisms such as convection also play their part. Finally, from the lower layers of the Sun's atmosphere, called the photosphere, light leaves the Sun. It is from this observable light that we estimate the chemical composition of the Sun. When passed through a prism, light emitted from the Sun is distributed on the basis of wavelength into the solar spectrum. This comprises a bright continuum of light with a complex series of dark bands. These 'absorption bands' are due to the absorption of light by cool atoms in the Sun's photosphere. The position of the bands in the spectrum tells us what elements are present; the intensity of the bands tells us how much of each element is present. However, this is an oversimplification.

To measure accurately the chemical composition of the photosphere we need information on its temperature and density, together with details of the conditions existing deeper down. All our data are imperfect, so it is no surprise that our knowledge of photosphere chemistry is only approximate. But even if our estimate of the chemical composition of the photosphere were precise, there is no compelling reason that it should be applicable to the whole Sun.

So, with all these 'ifs' and 'buts', we shall look at our estimate. In common with most stars, the Sun is composed dominantly of hydrogen and helium. These elements make up almost 99 per cent of the weight of the Sun. In contrast, the inner planets, Moon and meteorites contain very much less than one per cent of hydrogen and helium. It is a sobering thought that if the Earth formed from matter of the same composition as the Sun, the gas involved would have weighed about one hundred times the weight of the Earth now. If both gas and solid material had been present in a primitive Earth, it would have been a planet the size of Saturn, which is second to Jupiter alone. We believe that the Earth, Moon and other bodies of the inner solar system formed in a hydrogen- and helium-rich environment. At a later stage these gases were driven off by heat from within the solid bodies. Finally, radiation from the Sun expelled hydrogen and helium from the remainder of the solar system, unless they were trapped by a large gravity-field like Jupiter's. Compounds containing hydrogen such as methane and ammonia are, of course, abundant in the outer planets, and small amounts of hydrogen are still present on Venus, Earth and Mars, but in the inner planets the hydrogen is bound in sulphuric acid or water, which have high boiling points.

To compare the chemical composition of the Sun with the compositions of the inner planets, Moon and meteorites, it is therefore convenient to consider only the chemical elements heavier than hydrogen and helium. What we are doing is comparing the composition of about one per cent of the Sun with 99·9 per cent of inner planet, Moon and meteorite; and, with all the uncertainty in the estimate for the Sun, we find a fairly good match between the Sun and chondrites. In comparison, rocks at the Earth's surface are depleted in iron and nickel, and rocks at the Moon's surface are depleted in iron, nickel, sodium and potassium. This should not be surprising, for the Earth is still active and the Moon was active once. Within the Earth, iron and nickel probably became unevenly distributed between crust, mantle and core. The Moon, with its low gravity, probably lost sodium and potassium by evaporation to space. Some of the Moon's iron and nickel may be concentrated in a small core, but the body as a whole still seems to have lost these metals relative to chondrites and the Sun. The chondrites were never melted, so they have lost almost nothing of the solar system material from which they formed – almost nothing, that is, of the material that is solid at a temperature below a few hundreds of degrees centigrade.

In this discussion so far we have considered chondritic meteorites in general. But as described in the previous chapter there are three main groups of chondrites. The question naturally arises: Which chondritic group matches the Sun's composition best? The usual answer is that carbonaceous chondrites give the best fit, and, of them, CIs are normally chosen. The reason is that other chondrites are composed of both high and low temperature fractions, whereas CI carbonaceous chondrites are almost entirely low temperature mineral assemblages. Chondrules, which formed at high temperatures, have lost metals such as lead which evaporate at fairly low temperatures. The enstatite chondrites and ordinary chondrites are, therefore, depleted in volatile elements which appear to have been completely retained by the CIs. Thus the CIs are our choice as the best available sample of average solar system material which has lost only hydrogen, helium and other gases.

In Figure 5.1 the abundances of some 'heavy' elements in the Sun are plotted against their abundances in CI chondrites. Now, it is easier to measure ratios of elemental abundances in the Sun than to determine the absolute abundance of a particular element. Because of this, all other elements are plotted as the number of atoms relative to 1 million silicon atoms. This is a purely arbitrary choice. But in the study of meteorites, Moon and planets it is extremely useful, for it allows us to pick up trends in fractionation, for example, of iron relative to stony material. It must be noted that the scales in Figure 5.1 are logarithmic, going by factors of ten from 10 000 to 10 000 000. Points lying on the line with 1:1 slope have exactly the same measured abundances in the Sun and in CI carbonaceous chondrites, in each case relative to one million silicon atoms. Most of the elements plotted have abundances in CIs between a quarter and four times their measured abundances in the Sun. Considering the problems in performing a chemical analysis of the atmosphere of a star 148 million kilometres (92 million miles) away, this represents a good fit.

We can now proceed to compare the chemical compositions of different types of meteorite, Moon and planets with the composition of CI carbonaceous chondrites. Even if these last objects prove not to be representative of primary solar system material, they may still be taken as a base-line from which to measure compositional variation.

In Table 5.1 are presented the chemical compositions of three types of chondrite and a recent estimate of some chemical abundances in the Sun. The abundances of each of seven elements have been calculated relative to 10 000 silicon atoms. If we look at the figures for iron, we see that the Sun is apparently depleted in this element relative to CI; however, the estimation of iron in the solar photosphere has proved to be particularly difficult. Relative to silicon, the H- and L-groups of chondrite, too, have less iron than CI, the H, or high-iron, group being less impoverished than the L, or low-iron, group. Among the different meteorite groups we find that elements such as nickel and gold, which normally occur in metallic form, vary sympathetically with iron. A chondritic meteorite rich in iron is also rich in nickel and gold, and *vice versa*. The process that produced enrichment or depletion in iron relative to silicon has therefore been termed the metal–silicate fractionation. If we think back to what we know about the inner planets, we must realize how great was the importance of this process in establishing differences between the planets. Mercury has a high

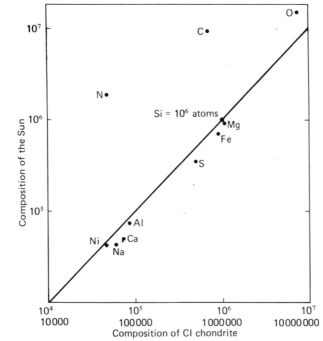

FIG 5.1    Abundances of the most abundant 'condensable' chemical elements in the Sun and CI chondrites, 'normalized' to 1 million silicon atoms. For explanation, see the text. Note the excellent agreement that is obtained, except for the three elements nitrogen, carbon and oxygen, all of which are volatile or form volatile compounds, e.g. carbon monoxide. As would be expected, these elements have higher relative abundances in the Sun, suggesting partial loss from CI and other planetary material of the inner solar system. (Data from Ross and Aller)

density indicative of the presence of a core making up two-thirds of the weight of the planet. Mercury, therefore, is highly enriched in iron and metal relative to silicon and stony minerals. On the other hand, the Moon is of low density and may not even have a core. In this body iron is inferred to be highly depleted – but more of the Moon later.

So, in the different chondritic groups we see evidence for the operation of a process that was probably fundamental in determining the properties of the inner planets. What was the process? There are various possibilities. If we think of the solar system when it was a rotating disc of gas and dust we can conceive of two mechanisms that may have favoured the adhesion of metal grains over stony particles. At low temperatures, magnetism of metal grains would encourage them to stick together. The second possibility is based on the malleability of iron-nickel metal; two colliding pieces of metal would tend to spread and

TABLE 5.1
*Chemical composition\* of H, L and CI chondrites, and the Sun*

|     | H      | L      | CI     | Sun    |
|-----|--------|--------|--------|--------|
| Si  | 10 000 | 10 000 | 10 000 | 10 000 |
| Ti  | 21     | 21     | 26     | 25     |
| Al  | 620    | 620    | 870    | 740    |
| Fe  | 8000   | 6200   | 8900   | 7100   |
| Mg  | 9500   | 9500   | 10 500 | 8900   |
| Ca  | 470    | 470    | 700    | 500    |
| Na  | 470    | 490    | 635    | 430    |
| K   | 29     | 28     | 40     | 32     |

\* Compositions are expressed as atoms per 10 000 silicon atoms.

adhere more than would pieces of brittle silicate. In either case, metal-rich bodies would have tended to grow faster than metal-poor ones. Both of these ideas have been with us for some time, but a third possibility was worked out more recently.

This theory begins by making the not undisputed assumption that the solar system formed as a hot nebula. If the composition of the nebula is known, then thermodynamic calculations can be used to specify the minerals that would have formed from the nebula during different stages of its cooling. The minerals thus formed are determined by the pressure and temperature of the nebular gas. So far, detailed calculations have been made only for a cooling gas of solar composition. It turns out that at pressures greater than about one ten-thousandth of an atmosphere, cooling solar nebular gas would condense iron-nickel metal before the onset of condensation of any abundant silicate such as olivine or pyroxene (FIG 5.2). Now, gas pressure in a discoid nebula is highest at the centre and lowest at the edge of the disc. The high pressure that must have prevailed in the central zone around the primitive Sun in the vicinity of Mercury's orbit would have ensured that here, metal would have condensed from the nebular gas, to form dust, long before olivine or pyroxene. The large core of this planet might therefore have aggregated from early formed metal grains before the condensation of abundant stony material had begun. Mercury and the other inner planets might have formed core-first, followed by the addition of silicate mantles after stony minerals had begun to condense. At the other extreme, very low gas pressure would have retarded the condensation of metal. In the outer reaches of the solar system, where pressures were low, olivine and pyroxene would have formed before metal. In this region, iron, nickel and gold would have become oxidized and so would have entered silicate minerals. At low temperatures, water would have reacted with silicates to produce the hydrous minerals found in CI chondrites. These meteorites may be associated with comets that are thought to originate in the outermost fringe of the solar system.

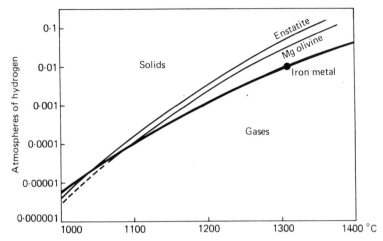

FIG 5.2    Condensation sequence in a gas of solar composition at different temperatures and pressures. Because the Sun is largely hydrogen, total pressure in the gas may be approximated by the partial pressure exerted by hydrogen. The pressure scale decreases in multiples of ten from one tenth of an atmosphere to one millionth. Note that for pressures greater than about one ten-thousandth of an atmosphere iron metal condenses before a magnesian olivine or enstatite (arrowed). Thus, in the inner part of a hot solar nebula, where pressures were high, on cooling, iron-nickel metal would condense at a temperature above 1100°C and before either of the most abundant silicate species. At lower pressures the major silicates would condense first. (Data from Grossman, 1972)

Thus, thermodynamic calculations based on cooling of a gas of solar composition can apparently account for the metal–silicate fractionation and for the chemical properties of various objects in the solar system.

If we turn again to Table 5.1 we can see that in H- and L-group chondrites aluminium, magnesium and calcium are depleted relative to CI carbonaceous chondrite; the Sun appears to be an intermediate case. These three elements normally form stony minerals and so are often referred to as 'lithophile'. By analogy with the metal–silicate fractionation, the process that produced differences in aluminium, magnesium and calcium, relative to silicon, between the different meteorite groups, is called the 'lithophile element fractionation'. Because uranium and thorium (also plutonium) are lithophile elements, the lithophile element fractionation must have played its part in determining the history of planets by its effect on the supply of radioactive, heat-producing elements.

Aluminium and calcium form various oxide and silicate minerals stable at high temperatures, this being less true for magnesium. It is this refractory quality that may account for the lithophile element fractionation. Once again, thermodynamic calculations have been made. Over a range of pressures the first material to condense from a cooling gas of solar composition is an oxide of aluminium. Such an oxide occurring at the Earth's surface would be known as corundum, or ruby, if bright red in colour. At a slightly lower temperature of perhaps 1200 °C the mineral gehlenite would condense. This is a calcium-aluminium silicate (formula $Ca_2Al_2SiO_7$) which, with falling temperature accepts more silicon and magnesium at the expense of aluminium. At lower temperatures still, the oxide mineral, spinel, becomes stable. Spinel is composed of aluminium, magnesium and oxygen only. Thus, at temperatures above the limits of stability of olivine or pyroxene a suite of aluminium, calcium and magnesium-bearing minerals would be formed. The lithophile element fractionation is thought to have been brought about by the condensation and segregation of these high temperature minerals. Planetary bodies that formed early may have been enriched in high temperature minerals, whereas later formed bodies may have lost out and are thus impoverished in aluminium, calcium, magnesium – and uranium and thorium.

In the late 1960s, irregular, whitish inclusions in CV3 meteorites, and a few rounded chondrules in CO3s, were found to contain the mineral assemblages predicted by thermodynamics for the earliest, high temperature condensates of a cooling solar nebula (see FIG 4.4 and Chapter 4). It is therefore possible that white inclusions represent the very stuff by which the lithophile element fractionation was effected. We return to them yet again in the next chapter.

Sodium and potassium are lithophile elements also, but, compared to aluminium, magnesium and calcium, they condense from a solar gas at lower temperature. Sodium and potassium are therefore grouped with other elements such as copper, silver and sulphur, into the 'moderately volatile elements'. From Table 5.1, H- and L-group chondrites have about a quarter less sodium and potassium, relative to silicon, than CI carbonaceous chondrites. To explain this it is usually argued that the parent-bodies of the ordinary chondrites captured only three-quarters of the dust that contained the chemical elements which condensed at about 700 °C. The remainder of the dust is presumed to have been lost to space or attracted to the Sun. Thus, the ordinary chondrites received only three-quarters of the available sodium and potassium and smaller fractions of other moderately volatile elements.

It must be emphasized that the various processes of condensation and accretion outlined here are no more than theories. An alternative hypothesis suggests that in the early solar system several planetary collisions occurred each with a large release of energy. This energy could have caused rocks to boil and lose matter by evaporation. The material left behind would have been enriched in refractories and metal. Deposition elsewhere of the material boiled off could have led to the formation of bodies enriched in volatiles. Planetary collision must have been highly improbable. But since we do not know whether planetary systems are common in the universe, we have no way of choosing between a nebular condensation model, which is required to explain a common phenomenon, or planetary collision which may have produced an all but unique solar

TABLE 5.2

*Estimates of bulk Earth composition, with CI for comparison*

|    | Earth I | Earth II | CI     |
|----|---------|----------|--------|
| Si | 10 000  | 10 000   | 10 000 |
| Ti | 17      | 42       | 26     |
| Al | 960     | 1280     | 870    |
| Fe | 11 100  | 12 500   | 8900   |
| Mg | 13 300  | 10 600   | 10 500 |
| Ca | 770     | 940      | 700    |
| Na | 126     | 134      | 635    |
| K  | 13      | 9        | 40     |

Elements in atoms per 10 000 silicon atoms.

Earth I: 68 per cent upper mantle, Table 3.2, plus 32 per cent core with 28 per cent of the Earth's iron, the remaining 4 per cent of the core being sulphur and nickel.

Earth II: Model of Ganapathy & Anders, based on the components of chondrites.

system. The strength of the collision theory is that it provides a source of energy to produce chemical fractionation among bodies born in a cold gas–dust cloud. Astronomical observation indicates that stars form in such regions and not in hot nebulae, hence astronomers tend to argue against the former existence of a hot solar nebula.

We have not yet looked at the chemical composition of the Earth in terms of meteorite chemistry, but before we do, let us briefly discuss how we may go about estimating the bulk composition of our planet. First, physical measurements indicate that the Earth has a dense core 32 per cent by weight of the whole Earth. We may infer by analogy with meteorites that this core contains 28 per cent of the planet's iron, the remainder of the core being nickel and sulphur. The other 68 per cent of the Earth comprise only the upper and lower mantle, for in global calculations like these we may safely neglect crust and oceans. If we *assume* that the upper mantle composition in Table 3.2, p. 46 is representative of the whole mantle, we can combine 68 per cent of this composition with 32 per cent core to derive the Earth I composition of Table 5.2.

A second approach is based more directly on our knowledge of meteorites. This is based on the assumption that the processes that controlled the fractionation of chemical elements during the formation of the Earth were the same as those in operation on the parent-bodies of chondrites. We can then use our observations on chondrites to derive a composition for the Earth. For example, we have just discussed the lithophile element fractionation in chondrites. In these, the elements aluminium, calcium and uranium are thought to belong to an early formed, high temperature condensate from a gas of solar composition. The ratios of aluminium to uranium, and calcium to uranium, appear to be constant in chondrites. Thus, if we can measure the Earth's uranium content we can calculate the aluminium and calcium contents from the appropriate ratios in meteorites. Now, uranium is the major heat-producer in the Earth, therefore if we can measure the amount of heat issuing from the planet we can calculate its average uranium content. One figure currently favoured is 0·018 ppm, or 18 grams (a little over $\frac{1}{2}$ ounce) per 1000 tonnes. Using inter-element ratios of chondrites and making an assumption about the oxidation state of iron, Ganapathy & Anders of the University of Chicago produced the Earth II composition of Table 5.2. However, it should be recalled that heat production in the Earth is still partly a mystery and so we may anticipate changes in the estimate of the bulk uranium content of the Earth. Such changes would necessitate modification of the calculated bulk chemical composition of our planet.

In spite of the entirely different approaches used in the two calculations, the bulk Earth compositions obtained are surprisingly similar. Both estimates indicate that the planet is enriched in metal relative to carbonaceous (or any other) chondrite. The Ganapathy & Anders, Earth II, composition shows an enrichment in titanium, aluminium and calcium, that is, in high temperature, refractory early condensate. The Earth I model has titanium depletion; this is explained by assuming that this element partly occurs in a sulphide-rich outer core and hence was missing from the upper mantle composition used as a basis for the calculation. In the Earth I model, magnesium is more strongly enriched compared with the slight enrichment in Earth II, relative to CI. The reason for the difference is that different magnesium to oxidized iron ratios were used in the two models. Unfortunately, physical data do not allow us to choose between the models, for the constraints imposed are wide. To a physicist, the lower mantle of the Earth may have the chemistry of an iron-rich pyroxene, or of a mixture of iron-poor olivine and pyroxene.

Finally, both calculations imply that the Earth is depleted in sodium and potassium by a factor of four relative to carbonaceous chondrite. We may safely conclude, then, that planet Earth is enriched in metal and high temperature refractories such as calcium, but depleted in more volatile elements such as sodium and potassium. Indeed, estimates of the abundance of the highly volatile metal – lead – indicate that the Earth lost over 90 per cent of this element relative to the CI abundance. But this loss is small compared with that undergone by the Moon. Our satellite is so depleted in volatiles that its primordial lead component is too small for its proper detection.

As outlined above, the estimation of the chemical composition of the whole Earth is largely speculative, the answer obtained being very much dependent on the prejudices of the individual scientist. But at least on Earth we have sampled bedrock down to 200 kilometres below the surface. This is certainly not the case with the Moon, for on none of the nine sampling missions was bedrock encountered. However, all nine missions returned various types of igneous rock of which the chemical compositions were established in the laboratory. From

TABLE 5.3
*Two estimates of the bulk composition of the Moon*

|  | Smith & Steele | Ganapathy & Anders |
|---|---|---|
| Si | 10 000 | 10 000 |
| Ti | 35 | 110 |
| Al | 720 | 3250 |
| Fe | 2630 | 2430 |
| Mg | 14 360 | 10 780 |
| Ca | 430 | 2400 |
| Na | 23 | 50 |
| K | 3 | 3 |

In each case, 5·5 per cent by weight of iron is metallic; the remainder of the elements listed above are combined with oxygen.

laboratory studies of the melting behaviour of rocks we can estimate roughly what solids must have been left behind at depth when lunar igneous rocks were produced as liquids.

Another approach is to assume that most of the upper part of the Moon melted and that the rocks at depth are olivine-rich residues left over from the melting. The travel characteristics of shock-waves associated with quakes on the Moon indicate that the top 60 kilometres are composed of plagioclase-rich crust, with basins containing basaltic rocks. Estimates of the compositions of crustal rocks, barren interior and a small, metal-rich core may be combined to yield a bulk composition for the Moon. This was done by Smith & Steele of the University of Chicago, and their estimate is presented in Table 5.3. Magnesium is the most abundant metal, as a consequence of the choice of olivine as the dominant mineral at depth. The mean density of the Moon – 3·34 – is consistent with an interior composed of magnesium-rich olivine.

Ganapathy & Anders applied the same reasoning as they used for Earth to obtain a chemical composition for the Moon. From physical considerations, the iron content was fixed at nine per cent by weight. The rate of flow of heat to the surface from depth was taken as indicative of a uranium content of 0·046 ppm, almost three times higher than on Earth. By assuming that uranium, titanium, aluminium and calcium were added to the Moon as the high temperature condensate of chondrites, the figure for uranium immediately fixed the abundances of the other three elements. In Table 5.3, it is evident that the Ganapathy & Anders' composition has titanium, aluminium and calcium contents three to five times greater than in the composition of Smith & Steele. It is interesting that both estimates stem from the same institution – the University of Chicago, but they are the products of different departments. In a discussion of their estimate of lunar composition, Ganapathy & Anders argue that melting in the interior of the Moon must have caused aluminium and calcium to have been concentrated upwards, for if high aluminium and calcium contents prevailed below about 600 kilometres beneath the surface, the high pressure mineral, garnet, would occur. Because of its high density, significant amounts of garnet cannot be present in the lunar interior, otherwise the bulk density would have been raised above the measured value of 3·34. It must be stressed that this estimate of lunar composition is critically dependent on the measurement of lunar heat-flow. Initially, a high value for the heat-flow was obtained, and linked to a lunar uranium content of 0·06 ppm. Later measurements of heat-flow suggested a lowering of the uranium figure to 0·046 ppm, which is quoted earlier in this chapter. Such a change in the estimate of lunar uranium content should have lowered the figures for aluminium and calcium contents by a quarter. However, the reduction in lunar uranium was almost exactly compensated by a reassessment by Ganapathy & Anders of the uranium content of the high temperature condensate in meteorites, which was lowered by a fifth. Thus a lower uranium figure in high temperature condensate was linked to the unchanged aluminium and calcium contents. So when the lunar uranium figure was reduced, estimates of the aluminium and calcium abundances stayed essentially the same. Because the lunar and meteorite measurements were carried out independently, there is no suggestion that the figures were manufactured to fit a preconceived model. It is hoped that the above example may illustrate some of the assumptions and uncertainties involved in the estimation of the bulk chemical composition of a celestial body.

We can safely conclude that, relative to Earth, the Moon is much poorer in iron and related elements, is poorer in sodium and potassium, and contains none of the highly volatile compound, water. But are we any closer to establishing the origin of the Moon in relation to the Earth? The answer is that we are not.

One certainty is that the Moon was in Earth orbit 3900 million years ago. Capture of the Moon would have been catastrophic, and we know that lunar basins and mountains had formed by this time. The Earth and Moon could have formed together, in orbit, as a double planet growing by accretion of condensed matter. If the 'seed' of the Earth were only slightly heavier than the 'seed' of the Moon, the small enhancement of the Earth's gravity would have caused it to capture more material than the Moon. The two would have grown at different rates, that for the Earth becoming ever faster as time went on, until all available condensable material was captured by one or other body, and growth ceased. This mechanism can explain some of the chemical differences between Earth and Moon. If the Moon had been unable to capture later formed, more volatile-rich material because of competition from the faster growing Earth, then the Moon should contain a higher proportion of early formed, high temperature condensate, as is the case. A fission hypothesis has its attraction, too. If the Moon had been broken away from Earth after the Earth's core had formed, this could explain the depletion of the Moon in iron and nickel, and in volatiles. A large body impacting on the Earth could have torn a chunk out of the Earth's mantle. The chunk would have been disrupted and heated, causing sodium, potassium, lead and other volatiles to boil off. The left-overs that ultimately formed the Moon would have been enriched in refractories, and the Moon's nickel, tungsten, gold, and much of its iron would have been retained by the Earth's core. However, the basis of this theory has recently been questioned. New experimental evidence indicates that a lunar core of less than five per cent by weight would be sufficient to accommodate all of the nickel, tungsten and gold scavenged from the remainder of the Moon.

Estimates of the bulk chemical compositions of Mercury, Venus and Mars are even more speculative than those of the Earth and Moon. The high density of Mercury indicates that it is iron-rich and oxygen-poor. This is apparently confirmed by study of the light reflected from the planet's surface, which limits the content of oxidized iron to less than 6 per cent. Comets, or other volatile-rich bodies, in elliptical orbits around the Sun are speeded up as they approach the Sun. Impacts on Mercury, the innermost planet, occur at high velocity with great release of energy. Because of its small gravity, Mercury is unable to hold volatiles brought to it from the outer solar system, hence explaining its oxygen-poor, metal-rich composition.

Russian measurements of the radioactivity at four sites on the Venusian surface indicate that the surface rocks are like basalt or granite. The planet, therefore, is differentiated. From its size and density, Venus is probably similar to Earth, though the Venusian core is likely to be smaller than Earth's. The dense atmosphere of mainly carbon dioxide is consistent with a bulk planet rich in oxygen. Temperatures at the surface are above those at which carbon dioxide reacts with silicates to form carbonate minerals, as in limestone on Earth. But if the temperature fell by less than $100\,^{\circ}C$, reaction would occur and atmospheric carbon dioxide would become locked in solid minerals. The Venusian 'greenhouse' would be destroyed, surface temperatures would fall, and a thin atmosphere dominated by nitrogen would remain. Thus, surface conditions on Venus might evolve towards those prevailing on Earth.

Although chemical analyses have been performed on surface materials at two sites on Mars, its internal constitution is just as speculative as that of Mercury. The high iron content of the 'soils', 13 per cent by weight, and the presence of an oxidizing atmosphere suggest that the upper parts of Mars contain more oxidized iron than the Earth's upper mantle. The partitioning of iron between oxidized mantle and metallic core of Mars means that the latter is probably smaller than the maximum of 19 per cent by weight. If Mars is more highly oxidized than Earth, it implies that the red planet contains a greater proportion of lower temperature, iron-bearing silicate than the Earth. Consequently we would expect Mars to have a smaller proportion of high temperature condensate than Earth. If this is so, Mars's uranium content must be less than the Earth's. Several problems are then posed, because a low uranium content would have resulted in heating insufficient for partial melting and early core-formation. Furthermore, without early core-formation and its associated upward concentration of uranium, more recent heating in the mantle would have been slight. That strong heating did occur is attested by the huge volcanic edifices on the planet. The source of heat for early melting on the Moon and Mars is one of the major problems in planetary science. It is a problem in meteoritics, too, and forms a recurring theme in the next chapter.

# 6
# What we learn from meteorites

In our discussion of iron meteorites it was stated that most belong to one or other of twelve chemical groups. This is generally taken to mean that iron meteorites are derived essentially from twelve planetary bodies. Irons that do not belong to any chemical group are called 'anomalous'. It is possible that some anomalous irons also are representative of different planetary bodies; so we can say that the Earth has received a sample from a *minimum* of twelve bodies. Because the meteorite sample arriving on Earth today is almost certainly the result of collisional break-up among the asteroids, time may be a crucial factor in determining our sample. This was discussed towards the end of Chapter 2. Anomalous irons may be the left-overs from a collision 1000 million years ago, most fragments from which have long since landed on a planet or else were sent along paths that took them out of the solar system. Alternatively, other anomalous meteorites may be the first arrivals on Earth from recent collisions, most of the fragments from which are stored in orbits that have not yet intersected the Earth's. It seems likely that discoveries made during missions to the asteroids (already under discussion) will include members of new chemical groups, so that the minimum number of planetary bodies represented by samples on Earth probably will rise from twelve.

There is no definite link between the five groups of chondrites and the groups of iron meteorites. The metal in enstatite chondrites has a chemical composition resembling Group IIAB irons. It is possible, then, that partial melting of a planetary body of enstatite chondrite bulk composition produced a IIAB core by draining of metal from silicate. The enstatite-rich residue may be the source of the enstatite achondrites, but there are arguments against this. In any case, we may conclude that the enstatite chondrites do not require the existence of a parent-body in addition to our minimum of twelve. These meteorites could presumably come from an upper, unmelted portion of the IIAB parent-body. The three groups of ordinary chondrites – H, L and LL – are more difficult to match with the irons. The H-group, like the stony-iron pallasites, may be related to Group IIIAB, but for the chondrites, the evidence is less compelling. The L- and LL-groups have metal, which, on average, is richer in nickel than H-group metal. The small amount of metal in LL-group meteorites has a high average nickel content – 20 per cent, or more. This would appear to indicate a similarity with IVB irons, but the LL-group metal is associated with sulphide and does not match the extreme depletion in volatile metals (e.g. arsenic and zinc) and sulphur exhibited by the IVB irons. Hence there is no apparent relationship between the LL or L chondritic groups and any of the twelve groups of irons. Most carbonaceous chondrites contain little or no metal and so, once again, there is no obvious link between them and the irons. The same is true of all achondrite groups excepting the enstatite achondrites discussed above. In spite of the generally negative evidence, we have to conclude that most meteorites may be derived from as few as twelve parent-bodies. The indications are that none of these was larger than Ceres, the largest asteroid and some 1000 kilometres in diameter.

### Meteorites come from bodies smaller than the Moon

Information on the size of a meteorite parent-body is of two different types. From laboratory experiment we can say that at high pressure certain minerals break down or else change from one type of crystal structure to another. If we find in a meteorite a mineral that we know is only stable above, say, a pressure of 40 kbar (40000 atmospheres), we could be sure that the meteorite came from a body the size of the Moon, or larger, because this is the pressure obtaining near the centre of the Moon, or about 120 kilometres down on Earth. The other type of approach is to establish how fast a meteorite cooled. If cooling had been rapid, we could say that the meteorite came from near the surface of a body from which heat was quickly radiating away. Slow cooling, on the other hand, would indicate deep burial in a body of a few hundreds of kilometres in diameter, or shallower burial in a larger body: remember, the interior of the Earth is still hot.

There are several minerals that may be used as pressure indicators. However, we must examine carefully the interrelationship of the various constituent minerals in a meteorite to ensure that a high pressure mineral was not formed in a brief period of high pressure induced by shock. By and large, this poses no problem, for the effects of shock are usually easily recognizable. When these are eliminated, we find that no meteorite contains a mineral assemblage formed above a pressure of about 12 kbar (12000 atmospheres) acting over a long period of time. Indeed, a number of meteorites of different types contain a high temperature, low pressure equivalent of quartz, which cannot form at a pressure as high as 3 kbar.

To emphasize these points further, it is convenient to examine some terrestrial rock-types. For example, near the Earth's surface, at low pressures, basaltic rocks consist essentially of the minerals plagioclase and pyroxene. The former contains all of the sodium and aluminium, and some of the calcium; the pyroxene contains most of the iron and magnesium, and the remainder of the calcium. But at a pressure of about 12 kbar and a temperature of 500 °C, basalt transforms to a rock called eclogite. Eclogite has a basaltic composition but is made up of the minerals pyroxene and garnet (Plate 4e). The pyroxene is a bright green variety and contains all of the sodium in the rock, but calcium, aluminium, magnesium and iron are distributed between both pyroxene and garnet. The result is that eclogite has a much higher density than basalt. A cubic metre of basalt weighing about 2·9 tonnes, when transformed into eclogite is compressed into 0·85 of a cubic metre; a cubic metre of eclogite weighs about 3·4 tonnes.

Eclogite is occasionally brought to the Earth's surface from depth, either in the roots of mountains formed by continental collisions (such as the Alps), or by kimberlites. Diamonds are sometimes found within blocks of eclogite brought up in South African kimberlite (Plate 4e)., If transport to the surface is fast and occurs with rapid cooling, the eclogite is unable to transform back to basalt. This allows it to survive on the Earth's surface in the same way as garnet-peridotite (Chapter 3). The main point of this argument is that plagioclase is unstable at high pressure. Sodium and aluminium from this mineral, must, at high pressure, find their way into alternative hosts.

When we look at chondrites, we find that *all* contain small amounts of plagioclase, although in a few cases it is present only as a component of glass. Furthermore, apart from the presence of metal and sulphide, the silicate portion of chondrites is not unlike olivine-nodules in chemical composition. We can therefore say that chondrites formed at pressures below 8 kbar, the lower limit at which the olivine-nodule mineral assemblage is stable. Some of the achondrite groups contain too little aluminium for plagioclase to form, but the eucrites have plagioclase in abundance. At high pressure these meteorites could reasonably be expected to transform to eclogite, but garnet is unknown in them. In fact, some contain tridymite, a high temperature, low pressure equivalent of quartz, indicating crystallization at a pressure of less than 3 kbar. Some of the mesosiderites and several types of iron meteorite with silicate inclusions contain tridymite, indicating that they, too, formed at low pressure. Recently, this conclusion was supported by a study of zinc-blende (zinc sulphide, or sphalerite) in some irons. The incorporation of iron into the zinc-blende crystal structure is pressure-dependent, and it was concluded that the small amounts of iron in this mineral are consistent with formation at pressures of less than a few kilobars.

Unfortunately, our evidence provides us only with upper limits to the pressures at which the various types of meteorite formed. This information cannot readily be equated with size of the meteorite parent-bodies. Tridymite is known in granitic rocks that crystallized within nine kilometres of the Earth's surface, this being the depth that is equivalent to the maximum pressure of 3 kbar which limits the stability of the mineral. Such a pressure would prevail at the centre of a body of chondritic composition (density 3·6) of 420 kilometres radius, approaching the size of the largest asteroid. Thus, mesosiderites and the few irons with tridymite could have come from depths as great as 420 kilometres. More likely, they are from shallower depths. Using the same argument, the chondrites must have formed at pressures below 8 kbar, the limit above which plagioclase breaks down, the component atoms going into spinel plus pyroxene. This pressure could be achieved at the centre of a small, chondritic planet of radius 700 kilometres, so this is the maximum depth from which a chondrite could have been derived. Measurements of the rates at which chondrites cooled from about 700 °C strongly suggest that these meteorites are from even smaller planetary bodies or from shallower depths than the limits set by pressure-dependent mineral assemblages.

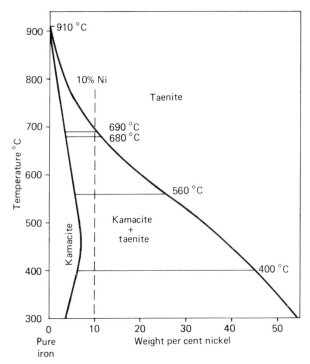

FIG 6.1   The iron-nickel system. The diagram is a means of expressing the relationship between temperature and the stability of compounds of different chemical composition, in this case, of different nickel contents. (See text) (From data of Goldstein and Ogilvie, and Romig and Goldstein.)

To establish the rates at which meteorites cooled, two different types of measurement have been applied. The first method dates from the mid-1960s, and depends on the rate at which nickel is partitioned between kamacite and taenite during the cooling of metal. Originally, the technique was applied to irons, but it was soon applied to the metal in chondrites and in stony-irons. The second method relies on the retention of different amounts of damage in different minerals brought about by the fission of plutonium atoms. The plutonium-fission technique has been applied mainly to chondrites, but its originator, Dr Pellas of the French Museum of Natural History, in Paris, is now extending it to other meteorites with suitable mineral assemblages, notably the silicate inclusions in IAB irons, and one pallasite.

To understand the method based on the partition of nickel between kamacite and taenite, we must glance at the diagram of the iron-nickel system (FIG 6.1). This merely illustrates the compositions - the nickel contents - found for the different minerals stable at different temperatures. Figure 6.1 shows that at temperatures above 910 °C and below the melting point, even pure iron has the taenite crystal structure. On and above the line running from 910 °C down to the right, taenite only is stable. The line represents the *minimum* nickel content that taenite must have at any given temperature. Kamacite is stable below and to the left of the line running steeply down from 910 °C. In this case the line indicates the *maximum* nickel content kamacite can have at any particular temperature. Between the two lines, under ideal conditions, iron-nickel metal consists of a mixture of both kamacite and taenite. Here, the compositions of kamacite and taenite are determined by the nickel content of the bulk metal, and by temperature. For example, metal with 10 per cent nickel by weight, at 560 °C, comprises kamacite with 6 per cent nickel and taenite with over 20 per cent nickel. Seventy-five per cent of the mixture will be kamacite, 25 per cent, taenite. At 400 °C and the same bulk nickel, the kamacite has about 6 per cent nickel and the taenite 45 per cent. About 90 per cent of the mixture is now kamacite. This illustrates that cooling of metal with a bulk composition found in many meteorites necessitates the growth of kamacite at the expense of taenite. If the meteorite is an iron, then the kamacite grows as four sets of plates parallel to the faces of an octahedron (FIG 4.11, p. 85) as the taenite shrinks.

With falling temperature, the supply of nickel pushed out by the advancing surface of kamacite exceeds the rate at which nickel can be incorporated into the adjacent taenite. This is because nickel atoms diffuse more readily through the kamacite structure than through the taenite structure at the same temperature. As temperature falls still further, the difference between the rates of diffusion of nickel in kamacite and in taenite becomes progressively greater. If cooling is extremely slow, nickel has time in which to diffuse into taenite. If cooling is fast, then nickel builds up along the margins of taenite, leaving the taenite interiors with a lower nickel content (FIG 6.2). It is this difference in the rate of diffusion of nickel through kamacite compared with taenite that allows us to calculate the cooling rate.

To estimate the cooling rate of an octahedrite, we need to know the bulk nickel content of the meteorite. This is done by chemical analysis; in fact, nowadays we can usually find such information in the literature. We need to know the thickness of the vestigial taenite into which nickel diffused from neighbouring kamacite. There is a geometrical method for measuring this on a polished and etched face, that is, using the Widmanstätten pattern. In addition, we must have an estimate for each of the diffusion rates of nickel through kamacite and taenite, for various temperatures. The diffusion rates are based on experimental measurements. Lastly, we must know the nickel content at the margin and centre of the residual taenite plates. This is normally measured using an instrument called an electron probe microanalyser, but new devices allow the electron microscope to be used. Once armed with these four pieces of information, the nickel profile observed in the taenite can be modelled using a computer, and the cooling rate adjusted until a good fit is obtained. However, it must be emphasized that the result is a cooling rate through a temperature of about 450 °C, the temperature around which nickel diffusion becomes sluggish. The method tells us nothing of cooling above 600 °C or below 300 °C.

Results show that many meteorites – irons, stony-irons and ordinary chondrites – cooled through 450° C at a rate between 1° and 10° per million years. The Group IVA irons are exceptional, cooling rates ranging from 5° to 100° per million years. This is compatible with the IVA irons having been dispersed through their parent-body, and with the variety of impact-produced features exhibited by members of the Group. If it is assumed that the parent-bodies of the meteorites had no prolonged source of heat and that the irons were from cores enveloped by stony material, we find that the maximum radius of the octahedrite bodies was about 200 kilometres. The chondrites, too, apparently came from bodies of similar size. However, one problem is that metal in unmetamorphosed ordinary chondrites such as Tieschitz (see Chapter 4) apparently cooled five times more slowly than the metal of strongly metamorphosed chondrites such as Kernouve. In Kernouve the cooling rate is determined as 10 C° per million years. One interpretation is that Tieschitz is from closer to the centre of its parent-body than Kernouve, and that the source of heat that metamorphosed the ordinary chondrites was external. Possibly, in the very distant past, the outsides of the planetary bodies were heated during an intensely radiative stage of the Sun. If intense solar heating were quickly 'switched off', then the interiors of small bodies would cool more slowly than the exteriors that had suffered higher peak temperatures. But this is pure speculation.

The above interpretation may not be valid for the ordinary chondrites. For the technique to be applicable, the distribution of nickel between kamacite and taenite must have taken place by diffusion in the solid state. In late 1979, this premise was questioned by Bevan & Axon, in consultation with the author. They found that in the unequilibrated (= unmetamorphosed) H-group chondrite, Tieschitz, metal and troilite often occur together in spherical blebs, like silicate chondrules. On polishing, then etching with acid, a structure was revealed in the metal (Plate 3c). Taenite was found to be composed of polygonal, dark-etching crystals surrounded by clear, high-nickel rims. This polygonal structure is analogous to the branched (dendritic) growth in silicate minerals that is brought about by the rapid cooling of chondrules of silicate liquid. Bevan & Axon, therefore, interpreted the structure of Tieschitz metal as essentially the product of a rapidly quenched mixture of metal and sulphide liquids, and not of solid state diffusion. The metallographic cooling rate of 1 C° per million years cannot, therefore, be applicable; the real cooling rate must have been very much faster. This work brings for the first time the interpretation of both metal and silicate into line: both cooled rapidly together as they agglomerated into the body that ultimately became the Tieschitz meteorite.

There is some evidence that metamorphosed H-group chondrites, too, cooled faster than their metallographic cooling rate – 10 C° per million years – would indicate. It thus appears that H-group chondrites are from a planetary body of

PLATE 1
a) Granite, near Saugues, Haute Loire, France. Note the coarse grain-size and the presence of cream coloured crystals of the potassium-rich mineral, microcline, some centimetres in length (see p. 41).

b) Olivine-basalt, Isle of Skye, in thin-section between crossed polars. Width of field 4·5 mm. In the petrological microscope for studying rocks in thin-section, white light from an illuminator may be polarized before it enters the section. Some minerals cause rotation of the plane of polarization, and when viewed through an upper polar which accepts light polarized at 90° to the lower one, a component of white light is seen, hence the variety of colour. Plagioclase occurs in elongate crystals with black, grey and white banding, which suggests a preferred orientation along the length of the field. Coloured crystals are mainly calcium-rich pyroxene (see Table 4.2), but olivine is also present (arrowed). Note that basalt is finer grained than granite (see p. 41).

c) Olivine-nodules in basalt, Sauterre, Puy de Dôme, France. The greenish, angular, coarse-grained blocks were brought to the surface by the basalt from 30 to 45 kilometres depth. The pale green mineral is olivine. A darker green mineral may be seen in the nodule, top right; this is a chromium-bearing pyroxene. Width of field 12 cm (see p. 43).

d) Kimberlite (blue ground) with diamond. The diamond is 1·5 cm across. This specimen, from Kimberley, South Africa, illustrates the fragmental nature of kimerlite; an angular fragment is arrowed (see p. 44).

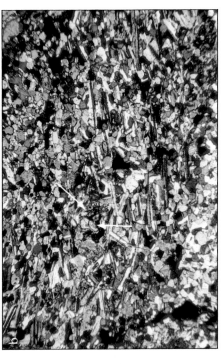

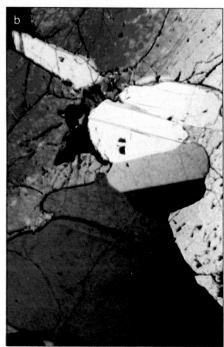

a) Garnet-peridotite block from the Kimberley Mine, South Africa. Width of field 8 cm. The specimen has been cut to reveal the texture. Note the buff coloured outer skin, due to hydrothermal alteration. Deep red, magnesium-rich garnets are conspicuous, as are crystals of green chromium-bearing pyroxene, e.g. in the buff coloured skin, top left (see p. 44).

b) Apollo XVII lunar highlands rock in thin-section between crossed polars. Width of field 1·6 mm. The rock is essentially composed of only plagioclase and olivine (orange-red-blue) and of a type called troctolite. ('Trout-stone', because of the speckled black and white appearance in terrestrial examples.) Here the plagioclase is calcium-rich anorthite. The rock was dated by the rubidium-strontium method at 4600 million years, but the interpretation is disputed (see p. 50). Courtesy NASA, Johnson Space Center, Houston.

c) Martian surface from Viking II, on the Utopian plains. The surface is strewn with rocks up to one metre across. Some rocks are not unlike terrestrial basalts in appearance, being dark and containing cavities' which, in basalts, form by the release of gases (bubbles) in the solidifying liquid. Winds have caused the red, fine-grained 'soil' to drift against boulders (see p. 58). Courtesy of NASA.

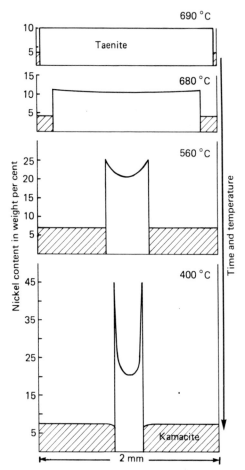

FIG 6.2    Development of 'M' shaped profiles of nickel in taenite. From Figure 6.1 we
see that on cooling to 690° C, metal with 10 per cent nickel enters the field of kamacite
+ taenite stability, causing kamacite to form as thin plates parallel to the octahedral
faces of the original taenite cubic crystal (FIG 4.11). Figure 6.2 illustrates the variation in
nickel content across taenite that has a newly formed kamacite plate on each side. By
680° C the taenite has become enriched in nickel to 12 per cent from the original 10 per
cent, but because the nickel abundance in the bulk is fixed, the taenite has to shrink in
size with a concomitant increase in the volume of kamacite. With further cooling,
kamacite continues to grow at the expense of taenite. However, diffusion of nickel
through taenite is slowing down, and nickel begins to build up at the boundary with
kamacite. The profile of the nickel content across the taenite now resembles the letter
'M'. Below about 450° C, the stability field of kamacite contracts to lower nickel contents,
so at 400° C kamacite is rejecting nickel which diffuses very sluggisly into the still

radius less than 100 kilometres and that, during cooling, Tieschitz resided at only 1 to 10 metres below the surface.

Metallographic cooling rates of irons and stony-irons have been criticized because the experimental evidence is based on simple systems that do not take into account the possible effects of minor and trace elements that naturally occur in the metal of meteorites. Although this is undoubtedly true, recent investigation of kamacite and taenite growth showed that the addition of the minor constituent, phosphorus, does not significantly affect the results. It therefore seems safe to conclude that most meteorites come from asteroid-sized bodies much smaller than the Moon.

When the solar system was formed, small amounts of plutonium and other unstable nuclear materials were present. We believe that these elements were blown out from an exploding star in which they had been synthesized by fusion processes, and a portion of them found its way into meteorites. The plutonium formerly in meteorites completely decayed with a half-life of 82 million years. This means that if, at some particular time, there were 1000 plutonium atoms in a mineral, after 82 million years only 500 would be left, the other half having changed into other elements by emitting $\alpha$ particles and, to a lesser extent, by fission. After a further 82 million years only 250 atoms would be left, and so on. After 10 half-lives, or 820 million years, had elapsed, only one atom would be left. At this stage we would consider that the plutonium was completely extinct. To measure the cooling rate in a meteorite we have somehow to relate fall in temperature with time; the rate of decay of plutonium gives us our clock.

When a plutonium atom fissions, it breaks into two, usually uneven, fragments. These fly apart with great energy. If a fragment enters a crystal, the atoms of the crystal are disturbed along a short path as the fragment is slowed down and stopped. The path of each fragment is therefore left as an invisible trail of damage. When the crystal is etched, perhaps with caustic soda, trails of damage passing through the surface are dissolved out. They appear as elongate holes running into the crystal in all directions, are known as fission-tracks, and are visible under the optical microscope. Heating, or prolonged storage at moderate temperature, allows the damaged crystals to repair themselves. This causes the fission-tracks to fade, or be annealed out. Fission-tracks are not registered in hot crystals, but during cooling, tracks begin to be retained at different temperatures specific for each mineral, giving us the means of measurement of temperature in our cooling meteorite. Tracks are retained at the highest temperature by plagioclase, followed by calcium-rich pyroxene, calcium-poor pyroxene, olivine and, at lowest temperature, by the calcium phosphate mineral, merrillite $(Ca_3(PO_4)_2)$. Now, in most chondrites merrillite was highly enriched

---

shrinking taenite. Enrichment in nickel at the margins of the taenite becomes extreme. Diffusion of nickel through kamacite is now slowing down, and nickel thrown out of the kamacite can not be replenished from within, hence the dips in nickel content in kamacite at its interfaces with taenite. (Based on the data of Wood, 1968)

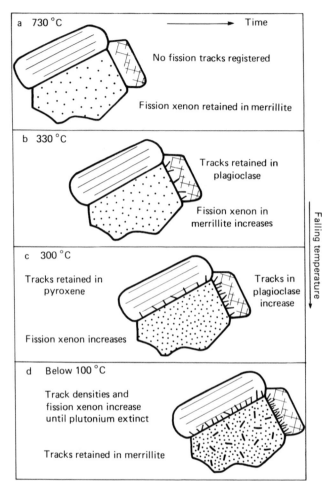

FIG 6.3   The record of plutonium fission in an ordinary chondrite. At very high temperature, fission products are lost and damage to crystals is repaired; there is no record of plutonium fission. Plutonium is concentrated in merrillite, a crystal of which is schematically shown in contact with pyroxene (upper left) and plagioclase (upper right).

  a) In an H-group chondrite, half the xenon gas produced by plutonium fission in merrillite is retained in the mineral as the temperature falls through 730° C. Xenon retention in merrillite is represented by stippling in the lower crystal.

  b) At 330° C tracks of fragments from fissioning plutonium atoms are retained in plagioclase crystals adjacent to merrillite. Tracks are represented by heavy lines from the merrillite interface into the crystal on the right. Xenon continues to be produced and to be retained within merrillite.

in plutonium relative to the other minerals. The technique employs merrillite crystals as reservoirs of plutonium, the adjacent crystals of different minerals being used as track-detectors (FIG 6.3).

The method is as follows: crystals of merrillite in contact with crystals of other minerals are polished and etched. Counts are made of the number of tracks in a given area of merrillite and in the adjacent minerals along their interfaces with the phosphate. Not all tracks result from plutonium fission, because some of the damage is caused by the natural fission of uranium, or by cosmic radiation. An estimate of the different contributions to the number of tracks must be subtracted from the total number of tracks in each crystal. The remainder may then be attributed to plutonium fission. The corrected track density in each mineral is then proportional to the amount of plutonium that remained in merrillite at the temperature of track retention. This is how temperature is related to time (FIG 6.4). (A correction has also to be made for track fading. Crystals of the different minerals are artificially irradiated with fission fragments in the laboratory. The crystals are then heated for set times at different temperatures. The effects of heat and time on artificial tracks are measured, allowing us to correct for fading in the naturally irradiated crystals.)

Results for L- and LL-groups of chondrites show good agreement with the metallographic cooling rates just discussed. For example, in the metamorphosed, LL-group, chondrite, St Séverin, the cooling rate from 650 °C to 280 °C was 4 C° per million years; from 280 °C to 30 °C it was about 1 ° per million years. The temperature of 30 °C is that at which all tracks in merrillite are retained. At 280 °C half of the tracks in low-calcium pyroxene are retained; the temperature of 50 per cent track retention is chosen for convenience because of the method for calculating the proportion of tracks that have faded. Every atom of plutonium that fissions produces two particles, each of which becomes an atom of another element. A proportion of the new atoms so produced are of the gas, xenon, which is like neon, but with much larger and heavier atoms. Atoms

c) At 300° C tracks begin to register in calcium-poor pyroxene (top left). They continue to register in plagioclase, and xenon continues to build up in merrillite.

d) At 100° C, tracks begin to register in merrillite. They continue to register in plagioclase and pyroxene; xenon continues to build up in merrillite. From the quantity of fission xenon in merrillite can be estimated the plutonium concentration in the mineral when the temperature passed through 730° C. From the number of tracks in plagioclase we can estimate the plutonium concentration in merrillite at 330° C. The difference in plutonium concentration between 730° C and 330° C is related to the half-life of plutonium, 82 million years. It was estimated that cooling between these temperatures took 50 million years for H6 meteorites, giving an average cooling rate between these temperatures of 8C° per million years. Differences in track density between plagioclase, pyroxene and merrillite give cooling rates between intervals at lower temperatures. Because plagioclase does not occur in large crystals in H5 and H4 chondrites (Table 4.3, p. 78) it cannot be used in estimations of their cooling rates.

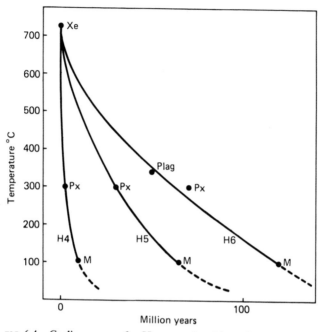

FIG 6.4    Cooling curves for H-group chondrites obtained by the plutonium fission method. 'Xe' is the temperature at which half the fission xenon is retained by merrillite. 'Plag' is the temperature at which half of the plutonium fission-tracks are recorded in plagioclase: 'Px' and 'M' at which half the tracks are recorded in calcium-poor pyroxene and merrillite. These temperatures are slightly higher than their equivalent temperatures given in the text because the H-group chondrites all cooled faster than St Sévérin. Figure 6.4 is an approximation based on the most recent (1981) data, which indicate that the parent-body of the H-group chondrites cooled faster and so was smaller than the parent-bodies of the L- and LL-groups. Note that H4s cooled faster than H5s, which, in turn, cooled faster than H6s, indicating that there is a stratigraphic relationship between the petrologic types. Type H4 came from nearer the surface of its parent-body than type H5, type H6 having been the most deeply buried.

of xenon from plutonium fission are tightly held within the merrillite crystals. Merrillite retains the xenon gas at temperatures far in excess of those that cause tracks to fade. Thus the quantity of fission-xenon in merrillite gives an indication of the plutonium abundance at a time before tracks were retained. On the graph, Figure 6.4, the 730 °C temperature at which half the fission-xenon is retained is used as the zero for our decaying plutonium clock. However, because of the great difficulty in physically separating enough merrillite crystals for measurement of the quantity of xenon, this cooling rate technique normally makes do with track-retention alone.

The number of tracks in each mineral adjacent to merrillite varies from chondrite to chondrite. This suggests that, if all merrillites originally had the same plutonium content – a reasonable assumption on the basis of chondrite chemistry – cooling began at different times. The zero time is, therefore, arbitrarily set, but in all cases represents a time at *about* 4500 million years ago. But although the different chondrites began cooling at different times, their rates of cooling were remarkably similar. However, in 1981 the rates for H-group chondrites were revised upwards, thus giving some support to the conclusion of Bevan & Axon.

Although the plutonium-fission cooling rate for St Séverin is lower than the metallographic cooling rate for similar chondrites, the agreement between the two methods is still good. Each technique gives a totally independent estimate and is subject to a variety of uncertainties. Also, each is applicable to cooling within its own particular temperature range. Finally, each indicates that metamorphosed ordinary chondrites come from the interiors, presumably outside the cores, of asteroidal-sized bodies no more than 300 kilometres in diameter.

There is also a third line of argument that favours asteroidal-sized bodies as the chief source of meteorites. Because meteorites exhibit the effects of cosmic irradiation, they must have existed in space as small bodies no more than one or two metres in radius. However, their textures and estimated cooling rates indicate that most meteorites come from the interiors of planetary bodies with radii between 50 and 200 kilometres. But the information is somewhat ambiguous and could apply equally to portions of a body the size of the Earth coming from within a few tens of kilometres of the surface. For a fragment to be removed from its parent-body, the fragment must undergo acceleration to a speed in excess of the escape velocity of its parent-body. In the case of a large body like Earth, acceleration to escape velocity would cause complete disintegration of material blown out by a major impact. Meteorites could not have come from a planet as large as our own.

The effect of a major impact on Earth may be seen in tektites. These are glassy objects the chemistry of which indicates that they originated on Earth. Tektites have been found in Georgia and Texas in the USA, in Czechoslovakia, in the Ivory Coast, and in Australia, the Philippines and south-east Asia. The four tektite fields are listed in order of decreasing age. Some Australian examples have structures indicating that they have undergone flight through the atmosphere at hypersonic velocities (FIG 6.5). Here we see a sample that originated near the Earth's surface and that was involved in an impact at an unknown location 700 000 years ago. The material was melted or vaporized as it was blown upwards. It then coagulated above the atmosphere and was ablated during hypersonic re-entry. No original structures have survived within them, yet the tektites could not even have reached escape velocity, but were returned whence they came.

Although the Moon has a lower escape velocity than the Earth, a lunar size still seems too big for the parent-bodies of meteorites. On the lunar surface,

FIG 6.5   Flight oriented tektite, Victoria, Australia. Width 2·5 cm. The rough, lower, surface was at the back during flight. Friction caused the front to melt and flow backwards as a flange around the unmelted core. Note the shiny surface on what was melted material.

glassy and broken rocks abound, and these are the materials that were unable to escape. In contrast, a number of meteorites have retained helium for over 4000 million years and in some unequilibrated ordinary chondrites delicate chondrules have remained in their pristine state. For many meteorites excavation from their parent-bodies must have been achieved with a minimum of violence. This precludes even lunar-sized bodies as the source of most meteorites.

## The solar system is at least 4500 million years old

From time to time in the text, the ages of rocks and meteorites have been mentioned. No technical details were given, apart from a simplified explanation, in Chapter 2, of an attempt at 'dating' the Earth using the lead content of uranium ore. But to comprehend fully this and the following section, some attention to detail is necessary; the reader lacking a knowledge of physics is referred to the Glossary.

First, let it be stated that the periods of time to be discussed are of the order of millions, or thousands of millions, of years. Because the experience of a single human normally lasts for about seventy years, longer time-spans must be viewed philosophically. Man, or something like him, has existed for perhaps three million years. Although this time is incomprehensible, we may still relate it to the much longer time that has elapsed since the Earth was formed.

The methods for 'absolute' age-determination utilize as their 'clock' the decay, at a fixed rate, of a parent radioactive isotope into a stable daughter isotope. For example, the first method discussed here is based on the decay of an isotope of rubidium to an isotope of strontium. Rubidium is an alkali metal and chemically similar to sodium, whereas strontium is an alkaline earth with properties akin to those of calcium. Because parent and daughter isotopes have different chemical properties, atoms of the daughter element have a tendency to migrate from their sites, in crystals, that had previously been occupied by atoms of the parent. Such migration is enhanced by the action of heat, which may cause the radiometric 'clock' to become reset.

One of the two isotopes of rubidium – $^{87}$Rb (rubidium-87) – decays to one of the four natural isotopes of strontium, $^{87}$Sr, with a half-life of approximately 49 000 million years. If, in a sample, we can measure the $^{87}$Rb content together with the $^{87}$Sr formed from rubidium decay, we can estimate the time since the sample formed, provided that neither rubidium nor strontium was added or removed. However, not all $^{87}$Sr is the result of rubidium decay. So in a single sample it is almost always impossible to measure the radiogenic $^{87}$Sr; 'radiogenic' means resulting from radioactive decay. We cannot readily distinguish the radiogenic from the 'common' component of the $^{87}$Sr. To obtain a reliable age-determination we need to find three or more rocks that we think have the same geological age but have different rubidium to strontium ratios. For each sample we must know the ratio of the number of atoms of the daughter isotope to the number of atoms of a non-radiogenic isotope of the same element. In this case we use the $^{87}$Sr/$^{86}$Sr ratio. We also must know the ratio of the number of atoms of the parent isotope to the number of atoms of the non-radiogenic isotope of the daughter element. That is, the $^{87}$Rb/$^{86}$Sr ratio. Isotopic ratios are measured in a machine known as a mass-spectrometer. The $^{87}$Sr/$^{86}$Sr ratios of the samples are plotted against their $^{87}$Rb/$^{86}$Sr ratios. If the points lie on a straight line, we know that the rocks formed at the same time and have the same age, which is proportional to the slope of the line. The line is called an 'isochron'. If the line is extended downwards to the left, it intersects the vertical axis at the point where $^{87}$Rb/$^{86}$Sr is zero. This defines the *initial* $^{87}$Sr/$^{86}$Sr ratio, that is, the uniform ratio shared by all the samples at the time when they became cool and when exchange of strontium between them ceased. A rubidium-strontium isochron plot is illustrated in Figure 6.6. All the points on the line with the steeper slope represent meteorite samples the $^{87}$Sr/$^{86}$Sr ratios of which have been increased over the initial value by the addition of radiogenic strontium.

Two examples of isochrons appear in Figure 6.6. The one with the steeper slope has the lower initial $^{87}$Sr/$^{86}$Sr ratio. This isochron shows that a selection of H-group chondrites cooled at the same time some 4500 million years ago. Because bulk samples were used, this is a 'whole-rock' isochron. A second type of isochron is called an internal isochron. If a rock or meteorite has remained

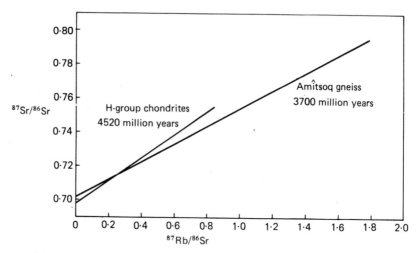

FIG 6.6   A rubidium-strontium isochron plot for H-group chondrites and Amîtsoq gneiss, the oldest rocks on Earth. Gneiss is a metamorphic rock, in this case of granitic composition, that has crystallized under pressure to produce a laminated texture. For the isochron, specimens of a number of meteorites were taken. The ratio of rubidium to strontium varies from sample to sample, thus allowing the method to be applied. The same is true for the gneiss, but here specimens were collected from a number of different locations. Note that the older isochron has the steeper slope and lower initial strontium isotope ratio of 0·6988. (Data from Minster and Allègre, and Moorbath *et al.*)

cold since crystallization, grains of each mineral act as if completely self-contained. Rubidium and strontium were bound into the structures of the different types of crystal when the rock was formed. Because different minerals are able to accept different, fixed amounts of rubidium and strontium, the $^{87}Rb/^{86}Sr$ ratio varies from one type of mineral to another. To obtain an internal isochron, a rock or meteorite is crushed to separate the mineral grains. Grains of each of its constituent minerals are then physically concentrated. In each mineral concentrate the $^{87}Sr/^{86}Sr$ and the $^{87}Rb/^{86}Sr$ ratios are measured. If there had been no mobility of rubidium and strontium since the rock crystallized, a straight line is obtained when the results are plotted on the isochron plot. Once again, the slope of the line is proportional to the age, and the initial ratio represents the $^{87}Sr/^{86}Sr$ ratio that obtained throughout the rock or meteorite when it had just cooled. In this way, the rubidium-strontium technique may be applied to a single, crystalline specimen, provided that its constituent minerals have a sufficiently wide range of $^{87}Rb/^{86}Sr$ ratios.

Members of the different groups of ordinary chondrites and of the enstatite chondrites formed some 4500 million years ago. This was originally shown by Wetherill and his co-workers at the University of California at Los Angeles. They produced a whole-meteorite isochron for each chemical group, indicating

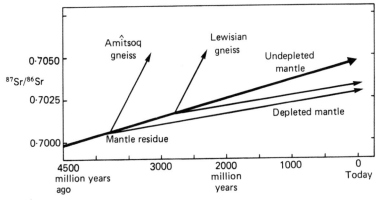

FIG 6.7   Schematic representation of the development of strontium isotope ratios on Earth. Today, basalts erupted from the mantle reservoir have ratios ranging from 0·7025 to about 0·705. Recent evidence indicates that the latter figure is representative of undepleted mantle. If initial ratios of mantle derived igneous rocks are plotted against time it is found that the Earth of 4500 million years ago had a strontium isotope ratio close to 0·699, as in chondritic meteorites and eucrites (basaltic achondrites, see FIG 6.8). Granitic rocks have high rubidium to strontium ratios and high rubidium and strontium contents. After a granite leaves the upper mantle, the residual mantle is more depleted in rubidium than in strontium. The rubidium to strontium ratio in the depleted residue is lowered, and production of radiogenic strontium is reduced, leading to lower strontium isotope ratios in depleted mantle, with time. Lewisian gneiss is the oldest British group of rocks. It occurs in north-west Scotland and ages extend backwards in time to about 2900 million years.

that the parent-body or parent-bodies of each group cooled at essentially the same time. The radioactive 'clock' begins to function when rubidium and strontium cease to be mobile. Then, $^{87}Sr$ atoms formed by $^{87}Rb$ decay are not pushed out of rubidium sites in crystals. The radiogenic $^{87}Sr$ cannot migrate elsewhere, but begins to build up at a rate exactly matching rubidium decay. Before rubidium and strontium cease to be mobile, any radiogenic strontium becomes mixed with common strontium. All rock or mineral samples in a genetically related group initially have a uniform $^{87}Sr/^{86}Sr$ ratio, which gradually increases with time. But, on cooling to the temperature at which rubidium and strontium are fixed in their sites, the $^{87}Sr/^{86}Sr$ ratio becomes the initial ratio for the system. If material is drawn off from a well-mixed system, and allowed to cool elsewhere, the initial ratio of the cooled material is fixed. In contrast, the strontium isotope ratio of the remainder will continue to grow, provided that the temperature remains high. Batches of rock removed at different times from a well-mixed reservoir have initial $^{87}Sr/^{86}Sr$ ratios that increase with time of removal. If the $^{87}Rb/^{86}Sr$ ratio of the reservoir can be estimated, differences in initial $^{87}Sr/^{86}Sr$ ratios between batches of material removed from it allow us to estimate the time between removal of the batches (FIG 6.7).

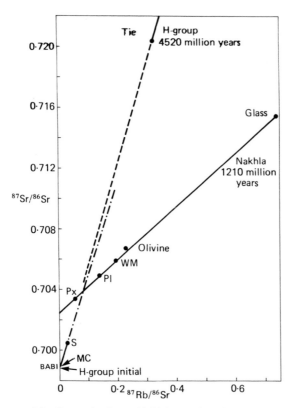

FIG 6.8    Determinations of initial strontium isotope ratios and isochron of the Nakhla achondrite. The eucrites have a very low rubidium to strontium ratio and little radiogenic $^{87}Sr$ has been added since they cooled about 4400 million years ago; their present strontium isotopic ratios are close to the initial ratios. The eucrite (basaltic achondrite) isochron, line S (Stannern) to MC (Moore County), although yielding a less precise age than the H-group chondrite isochron, does give a precise initial ratio. This ratio, termed BABI (see text), is within experimental error of the initial ratio of the H-group, indicating that the eucrites originally formed at the same time as the chondrites. The ordinary chondrites have high rubidium to strontium ratios and only Tieschitz plots in the field of the diagram. The Nakhla isochron is internal; Px pyroxene, Pl plagioclase, WM whole meteorite, olivine and glass were used. Note the initial ratio of 0.7025, which lies in the range of the Earth's upper mantle 1210 million years ago (see FIG 6.7). (Data of Minster and Allègre; Papanastassiou and Wasserburg; and Gale, Arden and Hutchison.)

Most of the achondrites have been shocked at different times after their crystallization and cooling. Transient reheating by shock produced a partial redistribution of rubidium and strontium that disturbed the radioactive 'clock'. A whole-meteorite isochron for the eucrites, or basaltic achondrites, yielded an imprecise age of 4400 million years. However, these meteorites contain almost no rubidium, but are rich in strontium. The whole meteorite isochron, by Papanastassiou & Wasserburg of the California Institute of Technology, yielded an initial $^{87}Sr/^{86}Sr$ ratio of 0·69899. This they termed BABI, an acronym from 'Basaltic Achondrite Best Initial'. Determined in 1969, it is still one of the most important numbers in the chronology of the Earth, Moon and meteorites.

At the time before planetary bodies began to form, rubidium and strontium were well mixed. Decay of $^{87}Rb$ to $^{87}Sr$ resulted in a steady increase in the $^{87}Sr/^{86}Sr$ ratio of the gas or dust from which the planets were to evolve. At some time more than 4500 million years ago a planetary body formed and melted. Either during condensation and accretion, or during melting, the moderately volatile rubidium was almost completely lost from the body. Strontium is refractory and was wholly retained. Because of their extreme paucity in rubidium, the basaltic achondrites, which were the melt-product, have had frozen into them the $^{87}Sr/^{86}Sr$ ratio of the solar system of over 4500 million years ago. This ratio is the BABI (FIG. 6.8).

Also shown in Figure 6.8 is an internal rubidium-strontium isochron for the Nakhla achondrite. The isochron shows that this meteorite crystallized 1210 million years ago, when the initial strontium isotope ratio, 0.7025, was in the same range as the initial ratios of many terrestrial igneous rocks. This indicates that both the Nakhla parent-body and the Earth have similar rubidium to strontium ratios. Together with other evidence, it then appears that Nakhla came from a planetary body that is chemically very like our own. Rocks 1200 million years old are not uncommon on Earth. They make up parts of the north-west of Scotland, of the Canadian Shield and of the African continent. But it is unlikely that Nakhla came from a body that is still active. To allow the meteorite to retain its primary igneous texture, it must have been broken from a body with a low gravity field. It seems unlikely that such a small body could have been internally hot only 1200 million years ago, when we know that the near surface rocks of the Moon were cold over 3000 million years ago. Therefore, we are forced to assume that Nakhla crystallized from a melt that was produced by a large impact on a small body. However, impacts are inefficient at producing melts, most of the energy being dissipated in fracturing and excavation, so we may yet have to seek a mechanism for releasing crystalline rocks from a large planetary body.

We have briefly examined the results of rubidium-strontium age determination on meteorites. There are at least three other methods of absolute age determination currently in use, but the story is essentially the same. However, because the uranium-lead method has implications for the Earth's age, we shall briefly consider some of the results.

Most uranium-lead ages of meteorites have been determined using lead isotopic ratios alone, without measuring uranium to lead ratios. This is possible because the method utilizes two different systems each of which has its own rate of decay:

$^{235}U$ decays to $^{207}Pb$, with a half-life of approximately 700 million years.

$^{238}U$ decays to $^{206}Pb$, with a half-life of approximately 4500 million years.

It is evident from the half-life that, in the solar system of 4500 million years ago, there must have been twice the present amount of $^{238}U$, but $^{235}U$ was then about 80 times more abundant than now. Then, decay of $^{235}U$ produced two atoms of $^{207}Pb$ for every atom of $^{206}Pb$ which resulted from $^{238}U$ decay. Today, $^{235}U$ is all but extinct, $^{238}U$ being 138 times more abundant. So, very little radiogenic $^{207}Pb$ is being produced now, while radiogenic $^{206}Pb$, from $^{238}U$, is still being produced at about half the rate of 4500 million years ago. The $^{207}Pb/$ $^{206}Pb$ ratio has therefore changed with time. A chondrite that formed 4500 million years ago and has since remained unaltered, will have retained all of the $^{207}Pb$ and $^{206}Pb$ generated since that time. It will have a high $^{207}Pb/^{206}Pb$ ratio. In contrast, bodies or masses of rock from which lead was removed during heating events in the past will have lost mainly the early formed $^{207}Pb$. Their present day $^{207}Pb/^{206}Pb$ ratios will be low. Thus, young rocks or meteorites tend to have lower lead isotope ratios than old ones. Also, the $^{207}Pb/^{206}Pb$ ratio of a rock or meteorite specifies the time since its formation.

In the early 1950s, some scientists in the USA, led by C. C. Patterson, began measuring lead isotopic ratios in meteorites. Lead typically occurs in sulphide minerals, whereas uranium has an affinity for oxide and silicate minerals. Thus, if stony minerals had been separated from sulphide some 4500 million years ago, and no further heating had occurred, lead in the sulphide would lack a radiogenic component. All the radiogenic lead would be associated with uranium in the stony minerals. In 1953, Patterson extracted lead from the troilite (iron sulphide) of an iron meteorite, and found it to have the lowest lead isotopic ratios measured at that time. Subsequently he measured the isotopic ratios of lead extracted from two chondrites and one basaltic achondrite. On an isochron plot, the three points from stony meteorites were found to lie roughly on a line passing through the composition of lead from the troilite. The slope of the line was consistent with an age of 4550 million years. In addition, it was established that the lead in recent terrestrial volcanic rocks, or in Pacific Ocean sediments, has isotopic ratios that lie on, or near, the meteorite isochron. It was therefore deduced that the Earth, too, has an age close to 4550 million years, a figure that has withstood the test of time.

A meteorite isochron is shown in Figure 6.9. This is known as a lead-lead isochron. The isotope ratio, $^{207}Pb/^{204}Pb$ is plotted against $^{206}Pb/^{204}Pb$; $^{204}Pb$ has no radioactive parent. Points lying on a straight line have the same $^{207}Pb/$ $^{206}Pb$ ratio, which is the slope of the line. The slope, in turn, is proportional to the age. The position of a point on the line is determined by the uranium to lead ratio in the sample. If the ratio is low, the lead isotopic ratios will be low because

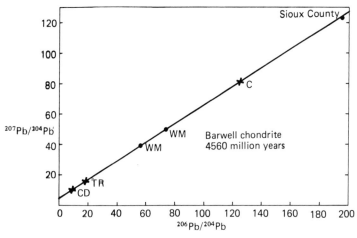

FIG 6.9    Modern version of Patterson's (lead-lead) meteorite isochron. Essentially an internal isochron of the Barwell (L6) chondrite (data of Unruh, Hutchison and Tatsumoto) using a troilite (TR) separate, two whole meteorite samples (WM) and separated chondrules (C), it is seen that lead extracted from the Cañon Diablo iron (CD) and from a whole meteorite sample of the Sioux County eucrite also plot on the line. The distribution of trace elements such as uranium and lead is notoriously inhomogeneous in ordinary chondrites, hence the different lead isotopic ratios in the two whole meteorite samples. Chondrules tend to be rich in uranium and poor in lead, hence the latter is very radiogenic, but not so radiogenic as the lead of eucrites. Note that the lead isotopic data indicate that Sioux County has an age close to 4560 million years, confirming that the eucrites and ordinary chondrites crystallized and cooled at about the same time, as deduced from initial strontium isotope ratios (FIG 6.8). Modern terrestrial lead plots close to the troilite point (TR), which is consistent with the Earth being about as old as most meteorites.

there will be only a small radiogenic component. On the other hand, a basaltic achondrite such as Sioux County has a low lead content and a high uranium content; its lead is highly radiogenic and it plots far up and to the right of troilite lead. Initial lead isotopic ratios cannot be obtained directly unless the uranium to lead ratio is known, and in the past it was usual to assume that the initial ratios were represented by those of lead from the troilite of iron meteorites. This is still largely true today. Although there are techniques for measuring uranium in the same sample in which lead isotopic ratios are determined, for chondrites the results are difficult to interpret. With colleagues from Oxford in England and Denver in the USA, I have been intermittently involved with this problem over the past ten years. It seems that either the initial lead of chondrites was richer in $^{206}$Pb than troilite lead from the Cañon Diablo iron, or else the lead in chondrites may have been contaminated. A further possibility was that the uranium in chondrites has a component that is isotopically different from the

PLATE 3

a) Cut and polished face of the Beddgelert, H5, ordinary chondrite, that fell in Wales in 1949. Width of field 7·5 cm. Note the even distribution of irregularly shaped, shiny metal grains, and the rather larger, grey to cream coloured, sub-circular chondrules (see p. 67).

b) Thin-section, between crossed polars (see caption, Plate 1b) of the Tieschitz, H3, chondrite. Chondrules are sharply defined, with dark rims, which are unusual. The yellow olivine crystal in the centre of the upper chondrule has irregular inclusions of dark glass indicative of rapid crystallization. The chondrule (lower, left) with grey to white crystals is composed dominantly of calcium-poor pyroxene. Field 1·5 mm. (see p. 67, 68)

c) Metal-sulphide chondrule in the Tieschitz, H3, meteorite. A polished surface seen through a reflected light microscope. Field 0·1 mm. Mid-brown: the iron sulphide, troilite. Pale: low-Ni metal – kamacite. Dark buff-brown, with margins: high-Ni metal – taenite. The section has been etched with acid which has preferentially attacked the interiors of taenite crystallites, which are dark brown. The margins of the crystallites are unattacked and appear buff, due to their higher nickel content of about 50 per cent. The chondrule probably originated as a molten droplet of metal plus sulphide. On cooling, nickel-rich metallic liquid separated as a droplet from sulphide-rich liquid. The metallic liquid crystallized rapidly to the oval area on the right (see p. 68). (Photograph by Dr S. O. Agrell)

d) Thin-section, between crossed polars, of Kernouve, H6. Note the granular, inter-grown texture and general absence of chondrules. A relic chondrule is visible in the centre of the field. It consists of radiating, grey-brown, fibrous crystals of pyroxene. Typical of the petrologic type (6), crystalline plagioclase is identifiable (banded, grey and black, arrowed on right). Width of field 4·5 mm (see p. 69).

e) Beddgelert, H5, in thin-section between crossed polars. No large crystals of plagio-clase are present, and chondrules are more prominent than in petrologic type 6. Some pyroxene crystals are striated, calcium-poor and intermediate in structure between the more disordered crystals in chondrites of petrologic type 3 and the ordered crystals in petrologic type 6. A striated pyroxene is arrowed (above and right of centre). Width of field 3·0 mm (see p. 70).

128

PLATE 4
a) A stone of the Orgueil, CI, metorite shower that fell in France in 1864. Width of field 10 cm. The dark fusion-crust is flaking away to reveal the interior which is a mixture of dark, water-bearing silicate and whitish carbonate and sulphate minerals. Carbon occurs also as organic compounds which are of interest in studies of the origin of life (see p. 71).

b) Cut face on a stone of the Johnstown, Colorado, hypersthene achondrite shower. Note the rounded fragments of green, calcium-poor pyroxene ('hypersthene'), which is chromium-bearing. The white groundmass is dominantly composed of comminuted pyroxene. Width of cut face, 10 cm (see p. 81).

c) Thin-section of Nakhla, a pyroxene-olivine cumulate, between crossed polars. An interlocking mass of calcium-rich pyroxene crystals, squarish to oblong in outline. Some are twinned, as indicated by change of colour across a diagonal. Slightly above and left of centre is a blue-purple olivine crystal. Top left (and elsewhere) are tiny plagioclases (elongate, grey to black). Width of field 3·0 mm (see p. 84).

d) Thiel Mountains pallasite. Brownish olivines are set in an enclosing network of iron-nickel metal. Contrast Figure 4.11, of the Crab Orchard mesosiderite, which is veined and fragmented. In the case of the pallasite, iron-nickel liquid was probably introduced into crystalline olivine, followed by slow cooling and crystallization of the metal. Width of field 7 cm (see p. 89).

e) Eclogite, with diamond, Newlands Mine, South Africa. On a cut surface, rounded, red garnets are seen in a groundmass of green, calcium-rich pyroxene. Rare diamonds are present (silver-grey). Width of field 2·0 cm (see p. 107).

Insert: An almost perfectly formed diamond octahedron close to an interface between garnet and pyroxene, on a weathered surface of the same specimen. The diamond is just less than 2 mm across.

uranium on Earth, but this was disproved recently. The answer is elusive, for this problem has been with us since the early 1960s.

The age of the Earth is only indirectly determined. It is based on assumptions, such as the choice of initial ratios of lead isotopes, which possibly vary from one planetary body to another. Therefore, it may be presumptuous to assume that a large body such as the Earth formed at exactly the same time as meteorites, and from materials identical to those from which the meteorite parent-bodies were built. But circumstantial evidence seems to indicate that this indeed was the case. Whatever the exact age of the Earth, the data from meteorites clearly show that the solar system was in existence over 4500 million years ago.

Earlier in the chapter, it was stated that plutonium fission-track studies indicate that different chondrites apparently began cooling at different times. So far, these time differences have not been confirmed by a method such as rubidium-strontium or uranium-lead, which is based on the decay of a long-lived radioactivity. In this case, we are measuring time backwards from the present day. On the contrary, cooling rates from plutonium fission utilize the decay to extinction of plutonium forwards from some unknown time in the past. Although the techniques overlap, we cannot calibrate one against the other. Thus there is apparently an unresolved conflict in that plutonium fission indicates that different ordinary chondrites cooled at different times, yet rubidium-strontium and uranium-lead ages of 4500 to 4550 million years encompass many chondrites. In the next chapter we shall return to plutonium and other short-lived, extinct radioactivities, for they provide information on their time of formation, in a massive star, relative to the birth of the solar system.

# 7
# Our legacy from the stars

Estimates of the age of the universe centre around 15 000 million years, give or take a few thousands of millions of years. A physicist's view appears to be elegantly simple. In the beginning, all matter was highly compressed. Then it expanded in a 'big bang' that caused a proportion of the primeval hydrogen nuclei to fuse into helium. Since then, the universe has been expanding and generations of stars have formed and died. Stars like our Sun obtain their energy as fusion reactors. At their centres, high pressure and temperature cause hydrogen nuclei to fuse into helium nuclei. The $^4$He nucleus, of two protons and two neutrons, is very stable. After the hydrogen fuel is consumed, if the star is massive enough to sustain in its core the higher temperatures and pressures required, $^4$He begins to be 'burned'.

Helium burning means the fusion of helium nuclei into heavier, stable nuclei. Because $^4$He is the basic unit, it is evident that the weights of heavier nuclei will tend to be in multiples of four. The one exception is $^8$Be (beryllium-8), which is unstable. When two $^4$He nuclei fuse to form the nucleus of $^8$Be, a third $^4$He nucleus must be quickly added before $^8$Be has time to decay – otherwise the process breaks down. On addition of the third $^4$He nucleus, the stable $^{12}$C (carbon-12) is formed in a proportion of cases. Thenceforward, a whole host of different nuclei may be formed, depending on the size of the star and the reactions involved. But the process stops when $^{56}$Fe (iron-56), the most stable nucleus, and associated nuclides are formed. Energy is released during elemental syntheses up to the iron group. When fusion ceases, energy production stops and the star shrinks under its own gravity. Small stars evolve to white dwarfs. Massive stars explode.

After the cessation of fusion, the iron-rich core of a star more than six times heavier than the Sun will collapse under its own gravity. Atoms are packed tightly together, forcing the nuclear structure to break down. The energy-release is collossal, and a supernova occurs (FIG 7.1). If the star is exceptionally massive, the core may collapse completely and a black hole results. Alternatively, the core may become a dense neutron star. Outside the core and before the supernova event, the massive star would have become layered. The innermost layer – or shell – in supernovae is iron-rich. The next shell is silicon-rich, and so on to generally lighter elements until the outermost shell of unburned hydrogen is reached. None of the shells is pure, but is composed of a dominant element, for example, carbon, with lesser amounts of other atoms, in this case oxygen, neon and magnesium. After the core collapses, it is thought that there is a rebound from which the release of energy heats the shells to a few thousands

FIG 7.1    The Crab Nebula. The remnant of a supernova of AD 1054 and some 6000 light years distant that was recorded by Chinese observers. The glowing matter is part of an envelope that is still expanding and carrying newly synthesized elements into interstellar space. Early in its history the solar system was seeded with similar material. It is even possible that the solar system owes its existence to the detonation of a supernova close to the gas–dust cloud from which it formed. The shock-wave from the impacting supernova remnant may have triggered the collapse of the gas–dust cloud, ultimately to form Sun and planets. Copyright by the California Institute of Technology and the Carnegie Institution of Washington. Reproduced by permission.

of millions of degrees centigrade as they are blown outwards. Protons, neutrons and $^4$He nuclei abound, and at the high temperatures that briefly prevail they may be added to nuclei to build the complete spectrum of the heavy elements which requires the input of energy. The explosion is not only responsible for atomic syntheses, but also causes the newly formed elements to become distributed in interstellar space. Ultimately the atoms may become incorporated into a star and planetary system, and into living bodies like our own. This is our legacy. It is our very being.

   Evidence that a supernova had showered the emerging solar system with newly formed atoms was discovered in a meteorite in 1960. The meteorite was the H-group chondrite known as Richardton, that fell as a number of stones in North Dakota in 1918. John H. Reynolds, of the University of California at

Berkeley, heated a sample of the meteorite under vacuum and extracted the xenon gas. The xenon was passed into a mass-spectrometer and the isotopic ratios measured. Reynolds found the isotope $^{129}$Xe to be particularly abundant; even when allowance was made for possible contamination by the Earth's atmosphere, there was still a $^{129}$Xe excess. This excess was attributed to the decay of $^{129}$I (iodine-129). The $^{129}$I nucleus is unstable, and tends to emit a $\beta$ particle. This has the effect of producing a proton from a neutron, and the $^{129}$I transforms to $^{129}$Xe. But the half-life of $^{129}$I is only 17 million years. This indicates that the $^{129}$I must have been synthesized shortly before it was incorporated into meteorites, otherwise it would have decayed to extinction, and the xenon gas would have been lost to space.

A refinement of the iodine-xenon technique was introduced in the late 1960s. Nowadays, before the xenon is extracted, the sample is put into a reactor and bombarded with neutrons. A known proportion of the $^{127}$I is thereby converted to atoms of $^{128}$Xe. The isotope $^{127}$I is the only stable isotope of the element and constitutes all of the iodine naturally existing today in meteorites and on Earth. After irradiation, the sample is placed in a furnace and heated under vacuum usually in 100° steps, beginning at 600 °C and ending at 1400 °C, or above. The gases released at each step are collected and separately passed into a mass-spectrometer. For each release temperature, the $^{129}$Xe/$^{128}$Xe ratio is measured. This is directly proportional to the $^{129}$Xe/$^{127}$I ratio. In general, at high temperatures in meteorites there is an excellent correlation between the release of $^{129}$Xe and the $^{128}$Xe produced by the irradiation of $^{127}$I. Therefore, we can be sure that the excess $^{129}$Xe really is from the *in situ* decay of an isotope of iodine – $^{129}$I. Otherwise we could not explain the observed relationship between the isotope of a gas, xenon, and a condensed element – iodine – in the minerals of meteorites.

The measured $^{129}$Xe/$^{127}$I ratio is identical to the $^{129}$I/$^{127}$I ratio that obtained when the meteorites studied originally cooled to the temperature at which xenon gas began to be retained. Now, in almost all meteorites, and on the Earth and Moon, the isotopic ratios of stable, heavy elements are constant. It is therefore safe to assume that when the solar system formed, the iodine everywhere had a uniform isotopic ratio. Differences in $^{129}$I/$^{127}$I between meteorites should therefore have a time significance. A meteorite from a parent-body that took a long time to cool should have a low $^{129}$I/$^{127}$I ratio, because most of the radiogenic $^{129}$Xe gas would have been driven off and lost to space. At the other extreme, a meteorite from a parent-body that formed and cooled early should have retained radiogenic $^{129}$Xe, so its calculated $^{129}$I/$^{127}$I ratio should be high. In practice, many chondrites and achondrites, and a few irons with silicate inclusions, all have $^{129}$I/$^{127}$I ratios within a factor of about two. This indicates that all of their parent-bodies cooled to below roughly 700 °C within approximately one half-life of $^{129}$I, or 17 million years (FIG 7.2).

Our argument may be taken further. The observed $^{129}$I/$^{127}$I ratios may be extrapolated backwards in time. This allows us to estimate the amount of time

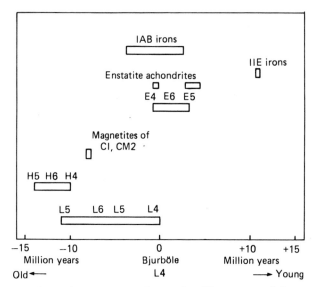

FIG 7.2    Iodine-xenon ages of meteorites. The ages are relative to an arbitrary standard, which is assumed to have a zero iodine-xenon age; the standard is the L4 chondrite, Bjurböle. Meteorites with radiogenic $^{129}$Xe to $^{127}$I ratios greater than in Bjurböle are older, and *vice versa*. Older ages are negative in the figure. Apart from the silicate inclusions in Group IIE irons, the meteorites or their mineral separates cooled to xenon retention temperature within a 17-million-year period. (From data compiled by Jordan, Kirsten and Richter, and data of Niemeyer.)

that could have elapsed between the formation of $^{129}$I in a supernova and the incorporation of the iodine into meteorite parent-bodies. When the parent-bodies of stony meteorites were cooling, the $^{129}$I/$^{127}$I ratio was close to 1 to 10 000. Ten half-lives, or 170 million years, before meteorite parent-bodies were cooling, the $^{129}$I/$^{127}$I ratio would have been about 1 to 10 (FIG. 7.3). Now, $^{129}$I is unstable, and although it could have been synthesized in a supernova in as great an abundance as the stable $^{127}$I, part of the solar system's $^{127}$I may have been synthesized in earlier events and incorporated into a pre-solar system gas-dust cloud before the formation of new $^{129}$I. Therefore, in the solar system, $^{129}$I could never have been as abundant as the stable isotope. The $^{129}$I/$^{127}$I ratio of 1 to 10 must be close to the highest possible limit. In turn, this sets an upper limit of 170 million years to the time that elapsed between synthesis of $^{129}$I and the cooling of meteorite parent-bodies. If we add a little time for uncertainties, we can safely conclude that some isotopes in the solar system formed in a stellar event less than 200 million years before the birth of planets.

The supernova in which $^{129}$I was synthesized could also have been responsible for the synthesis of a whole range of heavy elements. Among the likely ones are $^{244}$Pu and $^{235}$U. At least part of the longer lived $^{238}$U and $^{87}$Rb may

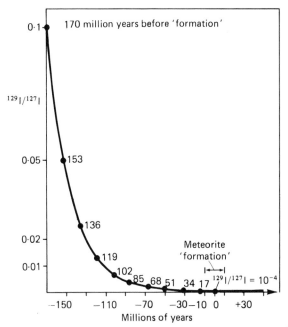

FIG 7.3   Estimation of the upper limit to the formation interval of $^{129}$I. The formation interval is the time between synthesis of the chemical element and its incorporation into the parent-body of a meteorite. The curve is the extrapolation backwards in time of the iodine isotopic ratio, subdivided into half-lives of $^{129}$I. (See text.)

have survived from earlier events. Thus, the inventory of the chemical elements heavier than helium has probably been built up over thousands of millions of years. Stable isotopes and long-lived radioactivities that formed in one supernova would have been dispersed into interstellar space. As interstellar dust they could subsequently become involved in a new generation of star formation, and so the cycle goes on. Our own Sun contains light and heavy elements from several periods of nucleosynthesis and is considered to be a third generation star. Short-lived radioactivities are the exceptions. They must have been produced immediately – in cosmic terms – before the formation of the solar system; and in 1974 was discovered evidence for an even more 'immediate' event than the synthesis of heavy elements up to 200 million years earlier.

   The CV3 group of meteorites of the carbonaceous chondrites contain centimetre-sized inclusions (Chapter 4). These comprise a suite of oxide and silicate minerals rich in the refractory elements calcium, aluminium and titanium. As previously stated, the mineralogy and chemistry of the whitish

FIG 7.4   An exceptionally large, coarse-grained inclusion in the Allende, CV3, meteorite. Width 2·4 cm. The minerals include calcium plagioclase (anorthite), gehlenite ($Ca_2Al_2SiO_7$) and calcium-rich pyroxene. Photomicrograph of a thin-section courtesy Dr R. S. Clarke, Jr, and the Smithsonian Institution, Washington DC.

inclusions were recognized in the late 1960s as those predicted for the first material to condense from a cooling gas of solar composition. In white inclusions, abnormally high levels of the isotope $^{26}Mg$ (magnesium-26) were discovered. The discovery was announced independently in 1974 by Gray & Compston of the Australian National University, and by Lee & Papanastassiou of the California Institute of Technology. This was exciting news for planetary scientists, because $^{26}Mg$ is the daughter isotope of $^{26}Al$ (aluminium-26), of which the half-life is only 740 000 years.

The discovery was to some extent brought about by chance. It just happened that in 1969 over 2 tonnes of stones of a CV3 meteorite fell in Mexico (Chapter 2). The Allende, Mexico, fall provided scientists with an abundance of material of what had previously been a rare meteorite type. Whereas in the past, research workers had to take great care to avoid destruction of irreplaceable meteorite samples, for the Allende fall this was not a major consideration. It meant that Allende samples could be dissected in the laboratory, and the white inclusions removed for separate study. This led to the finding, in 1973, of the presence of oxygen with a pure $^{16}O$ component. R. N. Clayton and his colleagues at the University of Chicago suggested that the $^{16}O$ had been introduced into a pre-solar nebula shortly before condensation began. The earliest material to con-dense – the Allende white inclusions – retained the signature of the high $^{16}O$ component. However, the bulk of this component was thoroughly mixed with the remainder of the gas and dust of the condensing solar system and so its separate identity was lost in many other solar system products. Well, if white inclusions had retained evidence of one isotopic anomaly, might there be others present? The search had begun.

The element, magnesium, has three stable isotopes – $^{24}Mg$, the most abun-dant, $^{25}Mg$ and $^{26}Mg$. The isotope, $^{26}Mg$, is not rare, and in a magnesium-rich mineral the addition of a small radiogenic component would be swamped by the non-radiogenic $^{26}Mg$ already there. Radiogenic $^{26}Mg$ was therefore sought in concentrates of the mineral, anorthite. This has the formula $CaAl_2Si_2O_8$, which you may recognize as one of the end-members of plagioclase (Table 4.2, p. 67) and which is an important constituent of the ancient lunar crust. Anorthite has a high aluminium content and almost no magnesium. So, had the radioactive isotope of aluminium been present, its daughter, $^{26}Mg$, would represent a significant addition to the minute amount of magnesium already present as an impurity in the mineral. This proved to be the case.

The different minerals in white inclusions (FIG 7.4) have a range of aluminium to magnesium ratios. When some inclusions are broken from the meteorite, gently crushed, and their minerals separated, the separates are found to have $^{26}Mg$ anomalies proportional to the aluminium to magnesium ratios. The higher the ratio in the separated mineral grains, the higher the excess $^{26}Mg$. A modified internal isochron plot (FIG 7.5) illustrates this point. Because $^{26}Al$ is completely extinct, $^{27}Al$, the only stable isotope of aluminium, must be substituted for it in the diagram. However, because the $^{26}Al/^{27}Al$ ratio in the different minerals was constant at any particular time, substitution of the stable isotope for the radioac-tive parent is valid.

The diagram indicates that the mineral assemblage crystallized and cooled when $^{26}Al$ was extant. Thus, $^{26}Mg$ atoms, formed from the decay of $^{26}Al$, were unable to escape from the original sites of the aluminium atoms and migrate to fill sites in crystals suited to magnesium atoms. The diagram also provides the $^{26}Al/^{27}Al$ ratio at the time when the minerals became closed to magnesium and aluminium exchange. Just like the $^{129}I/^{127}I$ ratios in other meteorites, the

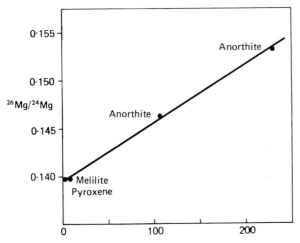

FIG 7.5 Aluminium-magnesium internal isochron from a calcium-aluminium-rich inclusion in the Leoville, CV3, chondrite. $^{24}$Mg is stable, with no radioactive parent and is the most abundant of the isotopes of magnesium. $^{27}$Al is the only stable isotope of the element. The isochron demonstrates that excess $^{26}$Mg correlates with aluminium content; that $^{26}$Mg is the daughter of $^{26}$Al. Anorthite ($CaAl_2Si_2O_8$) can have small amounts of magnesium as an impurtiy; two different mineral separates have different magnesium contents and hence different $^{27}$Al/$^{24}$Mg ratios. Melilite may be considered as gehlenite (FIG 7.4) in which magnesium and silicon replace some of the calcium and aluminium. From data of Stegmann and Begemann.

$^{26}$Al/$^{27}$Al ratios in Allende white inclusions have a time significance. But, because of its short half-life of only 740000 years, for $^{26}$Al in white inclusions we must think on a time-scale of only a few millions of years. The *initial* $^{26}$Al/$^{27}$Al ratio in some white inclusions was not much less than 1:10000. By analogy with $^{129}$I production, discussed earlier, we may be sure that the white inclusions had crystallized and cooled within a time-span of ten half-lives of $^{26}$Al from the time of its synthesis. That is, within 7 million years of some astrophysical event, the products had been incorporated into crystalline materials, now part of the solar system.

The interpretation of our data on calcium-aluminium-rich inclusions, sometimes referred to as CAIs, is not without dispute. Although the isochron relationship *proves* that the CAIs crystallized when $^{26}$Al was extant, there is no proof that crystallization took place in the solar system. It remains a possibility that CAIs originated outside the solar system, at some immeasurable time in the past. They may represent a generation of centimetre-sized objects introduced into a pre-solar gas-dust cloud. This interpretation is favoured by the astronomical observation that stars form in *cold* gas-dust clouds. If there never had been a hot solar nebula, then CAIs do not represent the earliest material to crystallize in the solar system. CAIs are, however, high temperature mineral assemblages, so how did they form? One possibility, put forward by D. D. Clayton of Rice University, in Houston, Texas, is that CAIs formed during the expansion and cooling of one of the outer shells of a supernova. This is conveniently close to the potential source to explain the presence of the various isotopic anomalies. It can also explain why there is generally no correlation between anomalies among the isotopes of heavy and light elements. During their expansion, the shells ejected from the supernova would have tended not to mix. Unfortunately, time of formation of the CAIs cannot be specified precisely.

But D. D. Clayton's ideas are not without their critics. Again from observation, the dust in interstellar clouds is of very much finer grain-size than the centimetre-sized CAIs. Also, it is argued that material in the expanding shells of a supernova could not condense and grow into sufficiently large chunks to match the size of the CAIs. If these criticisms are valid, it puts time and place of formation of the calcium-aluminium-rich inclusions fairly and squarely back into the early solar system.

A recent variant of these theories was proposed by O. K. Manuel, of the University of Missouri, Rolla. He suggests that the supernova was the Sun itself. The planets, he argues, have retained some of the chemical identity of the shells that were ejected. Thus, Mercury is iron-rich from the inner shells; Venus and Earth have both iron and stony material from intermediate shells; Mars and the asteroids are mainly stony; and the outer planets are helium and hydrogen enriched, from the two outside shells. In this interpretation, the Sun represents the supernova remnant that subsequently captured hydrogen from interstellar space and so recommenced hydrogen burning. Manuel bases his argument on the fact that isotopic anomalies in either light or heavy elements may be correlated in Allende inclusions. In contrast, there is no correlation between an anomaly in a light element such as oxygen and a heavy element such as xenon. (In this case, the anomalous isotopes of xenon do not include $^{129}$Xe.) Had the supernova occurred at some distance from the solar system, then, argues Manuel, the products should have become mixed during their journey. Because they did not mix, their journey must have been short and from the nearest star, the Sun.

In the more conventional view, $^{26}$Al was created in a supernova adjacent to the gas-dust cloud from which the solar system emerged. The time of synthesis of this isotope was within a few millions of years of the formation of the parent-body, or parent-bodies, of the Allende and related meteorites. So it is possible that the supernova acted as a trigger for the formation of the Sun and planets. The collapse of the gas-dust cloud from which the solar system formed could have been induced by compression from the shock-wave of the expanding supernova in which the $^{26}$Al was synthesized.

It now seems possible that $^{129}$I was synthesized in the same event as $^{26}$Al. G. J. Wasserburg and co-workers at the California Institute of Technology found that the abundance of $^{26}$Al relative to $^{27}$Al was rather less than 1 part in 10 000 at the time of cooling of many of the Allende white inclusions. The ratio of $^{129}$I to $^{127}$I when chondrites cooled to retain xenon had previously been shown to have a similar value. Third, in the past few years Wasserburg's team discovered that in refractory-rich, Group IVB iron meteorites, there is an excess of $^{107}$Ag (silver-107) which correlates with the abundance of the element palladium. (Palladium is a refractory precious metal not unlike platinum.) The scientists argue that the excess $^{107}$Ag is the daughter isotope of $^{107}$Pd (palladium-107) which has a half-life of 6·5 million years. Finally, the ratio of excess $^{107}$Ag (i.e. $^{107}$Pd) to the stable isotope, $^{108}$Pd, was found to be close to 1 to 100 000. It is highly improbable that chance was responsible for producing the similarity between the three isotopic ratios at the time of meteorite formation. The favoured alternative is that aluminium, palladium and iodine were synthesized together in a single supernova, and a small amount of its debris became mixed with the pre-existing gas-dust cloud that became the solar system.

The gas-dust cloud previously had received the products of an earlier, larger supernova that had synthesized very heavy elements such as uranium and plutonium. The $^{244}$Pu to $^{238}$U ratio in meteorites was very much larger than 1 to 100 000, so a separate event is required for their production. The original argument for the maximum time elapsed between iodine synthesis and the retention of radiogenic $^{129}$Xe in chondrites holds for plutonium and its fission xenon. It is therefore plausible that a pre-solar system gas-dust cloud received the products of a large supernova. Less than 200 million years later the gas-dust cloud was showered by the debris from a second, smaller supernova. The shock-wave from this second event perhaps triggered the collapse of the gas-dust cloud and so initiated the formation of the solar system within a time-frame of only a few millions of years. If planets, Moon and asteroids had accreted at about the same time as Allende, they too would have incorporated $^{26}$Al. Its decay would have released sufficient energy to melt any body of radius greater than a few kilometres, if the body contained 1 per cent or more of aluminium. This would be most convenient for planetary scientists.

One of the outstanding problems in planetary science is to determine the means by which asteroid-sized bodies could have been heated. The evidence indicates that many iron meteorites have been melted at temperatures above 1500 °C and most achondrites are igneous rocks. Also, chondrites are apparently from small bodies (Chapter 6), yet most were heated above, or slowly cooled through, 700 °C (Chapter 4). Moreover, absolute age determination dictates that this took place not less than 4500 million years ago. Possible sources of heat may be considered under four categories.

1. Radioactivities with half-lives as long as that of $^{235}$U – about 700 million years – release their energy slowly. In fact, small bodies of asteroid-size can radiate most of the heat from their surfaces, so that the temperature rise is only slight. If long-lived radioactivities had been the only source of energy, a larger

body such as the Moon would not have become internally hot until almost 1000 million years after its formation. Clearly, some other source of heat is required to account for the early formation of the lunar crust, and for melting in the parent-bodies of meteorites.

2. Bodies impacting on the surface of a growing planet release gravitational energy. Even if the Moon had formed in less than 20 000 years, the heat retained by it would have been sufficient to raise its *average* temperature by only 600 °C. For an asteroid, the gravitational energy is insignificant. Therefore, as the Moon grew by accretion an asteroid-sized kernel would have formed cold and most of the gravitational energy would have been deposited in the outer layers of the almost totally formed Moon. This would have led to melting in the outer few hundreds of kilometres while the deep interior stayed cold. On the other hand, a more massive body such as the Earth could have been completely melted by gravitional energy towards the end of accretion. Settling of molten metal into a core could have provided even more heat.

3. Large amounts of energy may have been released as radiation from an early, active Sun. Now we are getting 'hot'. From observation, some stars pass through a highly radiative stage before embarking on the main sequence of hydrogen burning, like today's Sun. The enhanced radiation may take the form of superluminosity – emission of intense light – or intense particle radiation – an extra-strong stellar wind. In either case, the energy release may be 10 000 times stronger than during the main sequence. Enhanced radiation may last for up to about one million years. Extra-strong sunlight could have heated the surfaces of the planets and asteroids, but more especially the inner planets. An enhanced solar wind, because it consists of electrically charged particles, would have caused electric currents to flow around planetary surfaces. Electrical heating would have ensued. This is doubly attractive, because magnetic effects would have been created in addition to heating, and the magnetism of meteorites may thus be explained. Finally, a word of caution. The early history of the Sun is unknown and it has not yet been established that it ever passed through a highly radiative phase.

4. Lastly, heating may have been caused by the decay of short-lived radioactivities such as $^{26}$Al. As already stated, this would have made an efficient heatsource. If the planets, Moon and asteroids had incorporated $^{26}$Al with the same abundance, relative to $^{27}$Al, as in calcium aluminium-rich inclusions, complete melting would have occurred within one million years of formation. The only exceptions would have been bodies of radius less than a few kilometres, diffuse bodies such as comets, or bodies with an abnormally low aluminium content. Only in small bodies that had never been hot would there still be evidence of a $^{26}$Mg anomaly. During the slow cooling of larger bodies, radiogenic $^{26}$Mg would have mixed with the abundant, non-radiogenic, $^{26}$Mg component. Thus, any radiogenic component would have become completely masked. It may therefore be inferred that the CAIs originated in very small bodies either within, or without, the solar system.

Although there is no 'hard' supporting evidence, it is worth mentioning here the possibility of the former existence of a superheavy element. In the late 1960s it was discovered that some carbonaceous chondrites are abnormally enriched in the heavy isotopes of xenon. This led some scientists to suggest that the heavy xenon is the product of fission of a nuclide heavier than $^{244}Pu$. The reason is that xenon from $^{244}Pu$ fission is less enriched in the heaviest isotope – $^{136}Xe$ – than carbonaceous chondrite xenon. Nuclear physicists found that atomic nuclei with more protons than the 94 in the plutonium nucleus are less stable than plutonium, but it was predicted that stability would again be approached in nuclei with 110 to 130 protons. The (as yet) purely hypothetical elements with 110 to 130 protons in their nuclei have been termed 'superheavy'. In Chapter 3 a superheavy element was mentioned in connection with lunar magnetism. During the course of writing this book, fission of superheavies has been invoked (erroneously in the author's opinion) to account for the distribution of trace elements in iron meteorites. Such is the pace of planetary science. In fact the most recent evidence all but disproves the possible existence of a superheavy element in iron and stony-iron meteorites. So superheavy fission may be considered as a potential, though unlikely, heat-source, without observational support.

At this stage of our continuing investigation of the solar system, we must conclude that either the decay of $^{26}Al$ or solar energy is the most likely cause of early planetary heating.

In addition to the chemical elements heavier than the hydrogen and helium, the stars also gave us chemical compounds of the types from which life presumably evolved. These compounds of carbon, hydrogen, oxygen and nitrogen occur in CI and CM2 carbonaceous chondrites. Similar compounds are inferred to be present as coatings on dust grains in interstellar space, and in the nebular envelopes around some stars. Compounds of carbon, some of which are fairly complex, are, therefore, ubiquitous. We shall now take a closer look at these, our most remote 'ancestors'.

In the search for the origin of life, three main lines of enquiry have been followed. The first looks for chemical fossils on Earth, and tries to identify the types of organism whose traces are present in the oldest sedimentary rocks. The second line investigates carbon compounds in meteorites, samples of the most ancient of solar system materials, together with their potential precursors – the interstellar grains. The last line is purely experimental. From simple starting materials such as water, ammonia and methane, attempts are made to synthesize carbon compounds in sterile conditions in the laboratory. The ultimate aim is, of course, to match the artificially synthesized materials with those encountered in the natural samples.

The investigation of the earliest forms of life on Earth is fraught with difficulty. Although we can be sure that the imprints in rocks of simple cells and algae testify to life on Earth some 3500 million years ago, we have been unable to detect evidence of life beyond this time. Moreover, in the most ancient of fossiliferous rocks in southern Africa and Western Australia, traces of organic chemicals cannot with certainty be attributed to the life-forms that existed at the time when the rocks were laid down. The reason is that it is always possible that the tiny amounts of organic traces are the result of more recent contamination. However, by studying the primitive organisms that have survived until the present, we can get some information on how life might have evolved from the simplest to the most complex of forms. Viruses do not have a cellular structure, but are merely groups of chemical compounds that only replicate by interacting with more complex, cellular, organisms. Modern viruses, therefore, cannot represent the earliest form of life. The simplest cells are found in the blue-green algae and in bacteria. These cells are called prokaryotes and have no proper nucleus within them. The blue-green algae are able to utilize sunlight as a source of energy. Most cells have well-defined nuclei and other structures. For example, the cells of green plants contain physically isolated concentrations of the substance chlorophyll, which utilizes sunlight to produce sugars and starch as a store of energy. Such structured cells are called eukaryotes. In terms of increasing complexity there is an apparent progression from viruses to organisms with prokaryotic cells to organisms with eukaryotic cells. Thus, it seems likely that virus-like, replicating chemical compounds were present on Earth before the blue-green cellular organisms of which we have the earliest fossil evidence. This however, is speculation.

The study of carbon compounds in meteorites has been largely one of identification. Because we know that many of the compounds of carbon in CI and CM2 carbonaceous chondrites are as old, or older than, the Earth, they must pre-date life on our planet. Contamination by modern organic compounds has bedevilled some of our investigations. In the 1960s, for example, traces of amino acids were found in the CI meteorite, Orgueil. The discovery was later discredited when it was shown that the quantity and identity of the amino acids were consistent with those from ten finger-prints! But less than a decade later, amino acids of abiotic origin were identified in the CM2, Murchison fall of September, 1969. That these amino acids were not formed by living terrestrial organisms was shown by their three-dimensional structure. Each molecule of an amino acid produced by organisms has the same three-dimensional structure as the next, because the genetic code does not allow for variation. In contrast, amino acids produced in laboratory experiments or occurring naturally in the meteorites have approximately equal proportions of molecules of each of the two types of three-dimensional configuration. Amino acids are the building-blocks of protein. Our evidence, then, tells us that some of the fragments from which proteins may be built were synthesized without the agency of life more than 4500 million years ago.

The simpler compounds of carbon are extracted from crushed carbonaceous chondrites by dissolution in organic solvents, such as mixtures of benzene and methyl alcohol. Additionally, carbonaceous chondrites contain a tarry substance in greater abundance than the simple compounds of carbon and which has proved extremely difficult to extract and identify. It has been called 'intractable polymer' in recognition of the problems encountered in its analysis. This is discussed in the next section which covers the experimental approach.

Experimental syntheses begin with the mixing in a clean, sterile flask, substances that are universally abundant and which contain carbon, hydrogen, oxygen and nitrogen – the most common elements in living cells. In some cases sulphur is included also. Energy is then applied. Ultraviolet light, electrical spark discharges, or simply, heat, are the forms of energy utilized in the experiments. Heat alone is not enough, for in this case a catalyst is needed to produce reaction. (A catalyst is a material that encourages chemicals to react, yet is not itself consumed in the reaction.) In the experiments in question, iron-nickel meteorite has proved to be a suitable catalyst. Alternatively, 'serpentine' minerals like those forming the bulk of CI meteorites may be used. The use of catalysts has been important industrially for making fuel oils from coal or natural gas, in, for example, the Fischer-Tropsch process. So, after ultraviolet light, or sparking or heating has been applied to the contents of our flask for anything from a day to six months, the energy is turned off. The flask will then contain a variety of compounds.

The simpler compounds dissolve in organic solvents, but the intractable polymer remains. When the tars are partially broken down by chemical attack, the fragments are found to consist largely of carbon atoms in six-membered or five-membered rings, some with nitrogen atoms. Many of the rings are linked to form complex structures and it is evident that the tars comprise molecules with many tens of carbon atoms. Because in the experiments energy was applied over a long period of time, the end-products tend to be stable against decomposition. For example, in one set of experiments carried out by Sagan & Khare of Cornell University, the quantity of ultraviolet light applied was greater than that which the materials would have received in interstellar space during the whole lifetime of our galaxy. Thus, we can predict that the carbon compounds experimentally synthesized are those that would survive in the hostile environment of interstellar space. So it should be no surprise that synthetic intractable polymer has properties like those of tarry substances in CI and CM2 chondrites. Moreover, synthetic polymer emits and absorbs radiation at wavelengths close to those of interstellar grains.

Further comparisons between artificially synthesized carbon compounds and those in meteorites are useful. In CI and CM2 chondrites the tars co-exist with carbonate minerals such as calcite. It was found that the ratio of $^{13}C$ (carbon-13) to $^{12}C$ is significantly greater in the carbonate than in the intractable polymer. This difference is matched by the products of Fischer-Tropsch experiments, but not by the products of experiments utilizing either ultraviolet light or spark-discharge as the energy input. Experimental evidence apparently favours the formation of carbon compounds in meteorites by the action of heat. However, conflicting evidence has just been obtained by three scientists at the

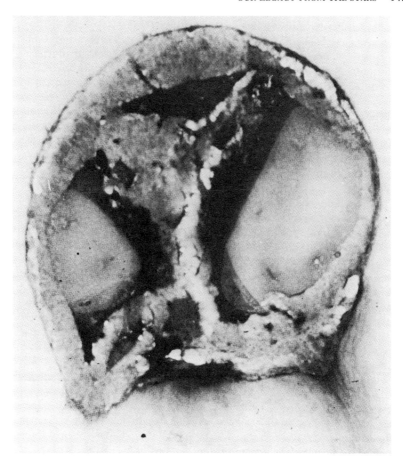

FIG 7.6  'Organized' element from the Orgueil, CI, chondrite. Electron micrograph, field 0·005 mm. Note the the suggestion of a cellular structure. From B. Nagy, *Geologiska Föreningens i Stockholm Förhandlingar*, 1966, volume 88, p. 259, Fig. 9, by kind permission of the Editor.

University of California at Los Angeles. They discovered that the hydrogen in carbon compounds in a number of carbonaceous chondrites is highly enriched in deuterium, the heavy stable isotope of hydrogen. This could could have been achieved by the formation of the compounds of carbon in the presence of water vapour at a temperature below 0 °C. Alternatively, enrichment in deuterium may be the signature of an interstellar component in the meteorites, because hydrogen in interstellar clouds often shows extreme enrichment in the heavier stable isotope. These are good illustrations of the different types of research now in progress.

In the past few decades it was occasionally suggested that 'organized elements' are present in some meteorites, but especially in Orgueil. One of the chief participants in these studies has been Dr. B. Nagy of the University of Arizona. Unfortuntely, such reports of extra-terrestrial life-forms have been largely discredited. The organized elements were either spherical or hexagonal cell-like structures about one thousandth of a millimetre in diameter (FIG 7·6). The hexagonal ones proved to be inorganic in origin, whereas some of the spherical ones are the result of contamination by spores or pollen when the meteorite was resident on a museum shelf. Although not every 'organized element' has been disproved as a pre-terrestrial life-form, there is now so much doubt involved that it seems likely that no undisputed evidence for extra-terrestrial life will be found in the meteorite.

If carbon compounds in CI and CM2 meteorites are older than life on Earth, where might they have formed? The best answer, but not the only one, appears to be that they were synthesized on dust-grains surrounding a star. Some stars of about the mass of the Sun, or larger, evolve to red giants after their hydrogen fuel has been consumed. Some red giants expel large quantities of energetic atoms of carbon, nitrogen and oxygen, and there is usually hydrogen about. These stars can, therefore, supply heat and materials from which carbon compounds may be synthesized on the less abundant silicate grains that act as catalyst. Radiation from these stars is notoriously unstable and may be powerful enough to strip off any envelope and distribute its contents in interstellar space. Subsequently, silicate grains with tarry coverings may become part of a gas-dust cloud, to become involved in a later episode of star formation.

Although the precise mode of formation of the tarry materials in meteorites is still a matter for discussion, the existence of such substances in interstellar space in universally accepted. The absorption and emission of light by grains in interstellar space are entirely consistent with the presence there of complex carbon compounds (FIG 7·7). The identity of the compounds is poorly known. Hoyle & Wickramasinghe argue that they are probably cellulose. However, it has previously been stated that 'intractable polymer' has not yet been properly identified in the laboratory. It therefore seems presumptuous to assume a precise identification of a compound on the basis of a few spectral lines in the light scattered by interstellar dust. This has been argued by Sagan & Khare, among others. (Cellulose is the material of which the walls of plant cells are built, and one of the most common organic compounds on Earth.)

As stated in Chapter 4, there is probably a relationship between CI and CM2 carbonaceous chondrites, meteors, and comets. Comets are thought to be aggregates of dust and ices formed in the outermost reaches of the solar system, beyond Pluto. Gravitational perturbations by stars or by Jupiter and Saturn can cause a comet to be deflected towards the inner part of the solar system. In this way, debris containing carbon compounds are showered on to the Moon and inner planets. Debris hitting the Moon's surface are completely vaporized, but some of the refractory elements such as platinum leave their traces in lunar

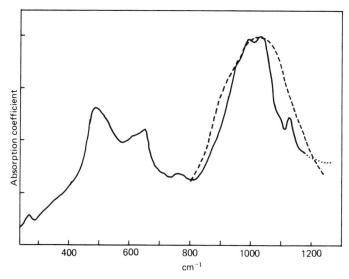

FIG 7.7    Absorption spectrum of finely powdered matrix of the Murchison, CM2, chondrite, compared with the absorption of interstellar grains (dashed line). Both have maximum absorption in this part of the spectrum at a wavelength of slightly less than one hundredth of a millimetre. Absorption at this wavelength is typical of fine-grained to amorphous silicates. Murchison does contain a variety of carbon compounds. It is argued that interstellar grains have silicate cores mantled by carbon compounds identifiable by an absorption peak at a wavelength shorter than that for the silicates. (From J. M. Penman, *Monthly Notices of the Royal Astronomical Society*, 1977, volume 175, p. 154, Fig. 5. By kind permission of the author and of the Royal Astronomical Society.)

'soils'. From this we estimate that the 'soils' contain about two per cent by weight of carbonaceous chondrite. Because of its greater gravity, our planet must receive an even higher flux of carbonaceous chondrite than the Moon; and on Earth, everything is retained, except, perhaps, some of the hydrogen. The same should also apply to Mars, where it was thought that superficial deposits might contain a significant component of carbon compounds. However, experiments on the Viking landers, although inconclusive, appear to argue against this. But we may still assume that the Earth receives a daily dose of primordial compounds of carbon. Moreover, there is no reason to believe that the flux of cometary material reaching Earth was less in the past than it is today. So we may be sure that when conditions on Earth became favourable for life, cometary messengers brought the necessary ingredients.

# 8
# Origin of the Earth and of life – a synthesis

Neither the age of the Earth nor its formation process is precisely known. However, we do know that our planet is part of the solar system, so presumably the Earth formed at much the same time as the Moon and meteorites. Lead isotopic ratios of several terrestrial materials are consistent with an Earth as old as meteorites, but this does not constitute absolute proof. Rarely, gases issuing from gas wells have an excess $^{129}$Xe. But again, this does not actually prove that the Earth formed around 4500 million years ago when $^{129}$I was extant. For alternatively, the $^{129}$Xe on Earth may have resulted from the decay of its short-lived progenitor in meteoritic material before it was added to the Earth during a late stage in its formation. However, the excess $^{129}$Xe from gas wells does prove that part of an ancient gas component is still present; that not all of the planet has been heated so severely that all of its internal, primordial gases were released. Our atmosphere, therefore, still receives an increment from within.

Let us assume, then, that the Earth originated over 4500 million years ago, and that its formation was linked to the formation of the solar system as a whole. But how might the solar system have formed? There are at least three possibilities. Unfortunately, because we do not know if stars with planetary systems are the rule or the exception, we are unable to choose which of the types of possibility is the most likely. If planar planetary systems are the rule, then the solar system almost certainly formed by a commonplace process. But if, on the other hand, stars rarely are central to planetary systems, our own star and its planets may have formed as the result of a highly improbable event, such as a close approach to our Sun by another star. Astronomical observation indicates that eight or ten near-by stars each has a satellite of some kind. Measurements, however, are not sufficiently refined to allow us to determine whether or not a planar planetary system is involved. Today we are tantalizingly close to answering the question of how many solar-type systems are likely to be situated in our immediate neighbourhood i.e. within about twenty light years. However, in the absence of a reliable answer we must conclude that none of the three possible origins for the solar system can be dismissed justifiably.

If planetary systems around stars are rare, it is conceivable that the Sun formed separately from the planets. This covers two of the three modes of origin. In the first type, the planets may have formed from matter pulled into space during a near collision between the Sun and a passing star. The second type of origin is one in which the planets formed from material captured by the Sun as it passed through an interstellar cloud. In the solar system, most of the

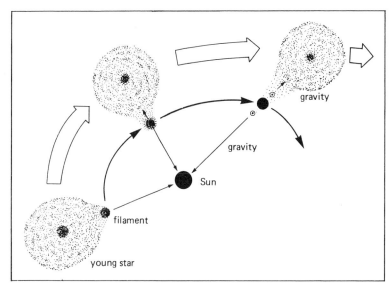

FIG 8.1   Representation of Woolfson's theory of the origin of the planets. A diffuse, young star approaches the more massive Sun. A filament is torn from the cool, young star and held where the Sun's gravity balances that of the young star. As the young star swings around and away from the Sun, angular momentum is imparted to the filament as it breaks up and condenses to form planets.

rotational energy (angular momentum) is lodged in the motions of the planets. Gravitational interaction between the Sun and a star that swept around it could have initiated the rotation of the material from which the planets formed (FIG 8.1). Although this appears to favour the first hypothesis, it is not compelling. The Sun has a powerful magnetic field. If the Sun had been spinning rapidly early in its history, its magnetic field-lines would have reached out like curved spokes of a wheel. This would have caused the planetary material to speed up, which, in turn, would have produced deceleration in the spin of the Sun, and so providing an alternative to stellar interaction to account for the angular momentum of the solar system.

The 1950s saw the decline of the catastrophic theory of the formation of the planets from a filament drawn from the Sun by a passing star, because a hot, gaseous filament from the Sun would expand and dissipate rather than condense into planets. In 1960, Professor Woolfson, now of York University, put forward an alternative, new theory. Stars tend to form in groups that are observed to contain stars of different ages. If an older star – the Sun – were approached by a young, cool star, then cool material drawn from the young star could have been held gravitationally for a time between the two stars. This could have allowed the filament enough time for it to condense into planets. As already stated, a hot filament would have dissipated. Later investigations by Woolfson showed that in the region of the Earth, only a single planet the size of Jupiter could have formed. To account for the inner planets, a second catastrophe was therefore necessary to break up the giant protoplanet into fragments, four of which became Mercury, Venus, Earth and Mars. This can also explain the slow, retrograde spin of Venus, which tends to be a stumbling block of non-catastrophic theories of the origin of the planets (FIG. 8.1).

In favour today are non-catastrophic theories in which Sun and planets are formed together. Descartes put forward one such theory in 1644, but here we shall briefly consider the 'floccule theory' of McCrea. This assumes that the solar system formed from part of an interstellar gas-cloud composed largely of molecular hydrogen. The hydrogen molecule comprises two atoms in co-existence. One important property is that molecular hydrogen is an efficient radiator of energy. Thus, when our hypothetical interstellar gas-cloud began to contract under gravity, the energy released was radiated away. In the most modern context, contraction of the cloud is thought to have been triggered by compression produced by the shock-wave from a near-by supernova. Associated with this event was the injection into the gas-cloud of newly synthesized $^{26}$Al, and an oxygen component highly enriched in $^{16}$O (Chapter 7). Heavier elements including palladium and iodine might also have been introduced. Contraction, with supersonic turbulence, caused the cloud to break up into a number of fragments, or floccules. Random motion caused floccules to coalesce, break up, and form again, until some twenty floccules came together to form the proto-sun. Ultimately, other floccules were brought into a single plane. Some began to rotate around the Sun, to become planets; the remainder were ejected from the solar system. At this stage, McCrea's theory is not unlike the modern version of that of Kant, who sugested in 1755 that a gas-filled universe with density variation would break up. Condensation would occur around centres of higher density.

In both versions, early condensed dust tended to settle to the equatorial plane of the original nebula, ultimately to form planetary cores. Outside a certain limit, centrifugal force prevented solid material from falling into the Sun, and solar radiation drove out the residual nebular gas from the inner solar system. Calculations indicate that collapse of a pre-solar gas-cloud could have taken less than one million years. This is consistent with the existence of $^{26}$Al in the early solar system.

So far, we have been discussing the origin of the Sun and planets in a general sense. Now we continue with a particular case, that of the Earth. In Chapter 5 we learned that the Earth is enriched in metallic iron-nickel but depleted in moderately volatile and highly volatile elements, relative to any group of chondritic meteorites. Today, the most commonly accepted theory of the formation of the Earth is one of heterogeneous accretion. That is, the composition of the matter that was added to the growing body changed with time. The first material to condense from a nebula of solar composition would have been refractory and

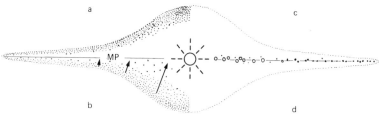

FIG 8.2   Representation of the origin of the solar system from a pre-solar gas-dust cloud, or solar nebula. As the nebula rotates it contracts because of the gravitational attraction between the particles; contraction may have been initiated by compression from the shock-wave from a supernova.

a) The nebula becomes a disc with a bulge around the centre. Material begins to condense.

b) Condensed material clumps into grains and falls to the nebular mid-plane (MP).

c) Grains aggregate to form larger bodies – planetesimals.

d) Solar radiation drives off residual gas from the inner solar system, as planetesimals grow into planets. At this stage growth is by capture, larger bodies growing at the expense of smaller. The asteroids are interpreted as the left-overs from this stage.

rich in aluminium and calcium. In the inner solar system this would have been followed by iron-nickel metal (FIG 5.2). So in the Earth's vicinity, dust falling to the nebular mid-plane would have been a mixture of metal and refractories (FIG 8.2). Because of magnetism or its adhesive properties, metallic dust would have clumped into small bodies more readily than the refractories (Chapter 5). Thus our planet may have grown core-first. Subsequently, when gravity increased, refractories and silicates would have been added to form the mantle. Finally, assemblages rich in hydrous minerals and carbonate, such as CI or CM2 carbonaceous chondrites, would have been last to accrete.

Current theory, supported by mathematical modelling suggests that the Earth was the largest of the swarm of planetary bodies that grew from dust in its neighbourhood in the inner solar system. As attested by the concordance among the ages of meteorites, growth of a small planet to asteroidal size was probably rapid. In contrast, growth of the embryo Earth by 'sweeping-up' asteroidal-sized bodies probably took about another hundred million years. The Earth, therefore, may be regarded as more youthful than the asteroids which, in part, probably represent the left-overs from the original swarm of small planets.

The alternative to heterogeneous accretion is homogeneous accretion. This theory of the accumulation of the Earth from dust of uniform chemical composition, although less popular now, does have its advantages and its supporters. If stars do not form in hot nebulae, then the condensation sequence outlined above could not have taken place. The planets must then have formed from interstellar grains, or something similar, with low temperature, water-bearing minerals. However, even if the process of accretion had been slow and the Earth actually formed cold, perhaps as part of a giant proto-planet, we know from the

Moon and meteorites that planetary bodies in the solar system suffered a period of early, intense heating. The Earth, by virtue of its large size, must have been even more strongly heated than the Moon. Our planet must have been heated so intensely that nearly all of the gases in its interior would have been driven off. Most oxidized iron would have been reduced to metal, as in a blast furnace, and would have settled to form or to enlarge the core. Thus the question of a hot or cold origin is of no real consequence for the solid Earth. For the formation of the atmosphere and for the origin of life it is absolutely crucial.

If the early atmosphere had formed by the release of gases from a hot, metal-bearing interior it would have been dominated by hydrogen, water, methane and carbon monoxide, and probably contained ammonia rather than nitrogen. This is an oxygen-poor, highly reducing atmosphere. Recent measurements of the oxidation state of samples from the upper mantle indicate that it is much less oxidized than the crust and hence lends some support to the theory of a reducing primitive atmosphere. Such an atmosphere has been used in experimental syntheses of a variety of compounds of carbon and seems to provide an easier pathway to life than more oxidizing alternatives. Had the Earth formed by heterogeneous accretion, the primitive atmosphere would have been comprised of gases issuing from low temperature, metal-free, near-surface rocks. Carbon dioxide, water and nitrogen are the most likely major components. This is a non-reducing atmosphere from which syntheses of organic compounds tend to be slow, although experimentation with potential catalysts is in its infancy. Neither type of atmosphere contains free oxygen, which is thought to have been produced by plants over thousands of millions of years.

After a few hundred million years the early histories of the Earth and Moon must have diverged. Early melting on our satellite produced an anorthosite layer some 60 kilometres in thickness. On Earth, however, an anorthosite veneer could not have been substantially thicker than 30 kilometres. The reason is that with the higher gravity and, therefore, steeper pressure-gradient on Earth, the high pressure mineral, garnet, would have replaced anorthite below this depth. Because garnet is dense, garnet-bearing rocks crystallizing below 30 kilometres depth would have tended to sink. So the early crust of the Earth was probably thinner than that on the Moon, and therefore would have been prone to fracture. Also, if garnet-rich rocks stuck to the underside of the crust, their high density could have caused whole slabs to founder into the underlying 'magma-ocean'. Moreover, the thin crust of the Earth would have been more readily penetrable by impacting bodies than its lunar equivalent. It is possible that events during this early period were directly responsible for initiating the process of ocean-floor spreading. We may conclude, then, that the Earth had a less stable, more turbulent early history than the Moon.

Following melting and in the absence of an atmosphere (the primitive atmosphere would have been driven off by the heat), radiation from the surface would have caused a solid crust to form on the Earth. This may have been the case 4400 million years ago. But, as argued above, the crust must have been thin, hot, and easily broken and penetrated by underlying molten rock. Again by analogy with the Moon, the Earth must have been heavily bombarded by bodies left over after formation of the planets. A proportion of these bodies would have been carbonaceous and water-bearing. This is borne out by lunar 'soils', which contain up to about two per cent by weight of dehydrated, volatile-free residues of carbonaceous chondrite. The Earth's upper mantle today contains two-tenths of one per cent of nickel. In contrast, stony materials in metal-bearing meteorites contain only two-thousandths of one per cent. So, unless very special circumstances are invoked, our planet must have received a late addition of nickel (by heterogeneous accretion) after the core had formed. To account for nickel in the upper mantle of the Earth, we can assume the late addition of carbonaceous chondrite material weighing, in all, two to three times more than the Moon. This could also have been the source of all of the Earth's sodium. The theory is, however, not without its problems. Compared to carbonaceous chondrite, the upper parts of the Earth are depleted in sulphur and lead, so we are forced to assume that these elements either evaporated and were lost to space, or melted and were added to the core. In either case, addition of carbonaceous chondrite must largely have taken place when the Earth's surface was still hot, but during a period of protracted cooling so that, ultimately, water and other gases were retained.

As time went on, cooling continued until the temperatuure of the Earth's surface fell to between 373 and 100 °C. The higher figure is the critical temperature of water and in this temperature range the Earth probably experienced its first rains and formed its earliest oceans. We do not know precisely when this occurred, but it was not later than 3800 million years ago, when the earliest sediments were laid down in liquid water. Early rain probably was corrosive, being laden with chlorine and fluorine or their compounds such as hydrogen chloride (hydrochloric acid gas). These gases, mainly introduced as part of the carbonaceous chondrite component, are still released by volcanoes today. The corrosive, acid rains would have dissoved sodium and potassium from rocks and added them to the oceans. As chlorine became bound in dissolved salt, the atmosphere would have become less acidic. As pointed out in Chapter 4, carbonaceous chondrites are probably linked with comets, the spectra of which indicate that they contain various compounds of carbon, hydrogen, oxygen and nitrogen. These must also have made up a proportion of the atmosphere of the juvenile Earth. All the evidence argues against the presence of unbound oxygen, so no ozone layer could have been in existence. Energetic ultraviolet light would not have been screened, but would have fallen on the surfaces of oceans and early continents, if the latter existed. On Earth, the scene was set for the beginning of life.

We have still much to learn about the road along which abiotic compounds evolved to become life. Had the atmosphere been non–reducing, as suggested in the preceding paragraph, then extra steps may have been required to produce ammonia or methane for organic synthesis. However, rocks of the crust or upper mantle seem to be able to act as reducing agents today. The lower of the two oxidation states of iron possibly plays its part in producing methane from carbon dioxide and water. Methane and hydrogen certainly issue from wells and volcanoes. Even if syntheses of carbon compounds are somewhat inhibited by a non-reducing atmosphere, it should be noted that some syntheses are possibly occurring now in brines fed by juvenile waters in the depths of the Red Sea. A further possiblity is that ultraviolet light could have caused water to decompose, releasing hydrogen to the atmosphere and leaving the water slightly enriched in oxygen. Thus, the water-atmosphere interface would have been a location for chemical reaction. Hydrated silicates would have been conveniently available to catalyse reactions.

Hoyle & Wickramasinghe have taken the opposing view that life originated in the centres of comets, which were gently heated by chemical reaction to temperatures above the melting point of water. They further argue that viruses such as influenza may still reach the Earth from comets. The more conventional argument, expressed by Sagan & Khare, is that viruses and their cellular host-organisms must be interrelated, so all originated and evolved here on Earth.

Whatever the pathway, it is likely that many and varied carbon compounds, including amino acids, were available on the prebiotic Earth, either from syntheses on the planetary surface or via meteoritic carriers. Amino acids are characterized by the presence of amino ($NH_2$) and carboxyl (COOH) groups. Two amino acid molecules may be bonded by the elimination of a water molecule ($H_2O$), the H derived from the amino, the OH from the carboxyl groups. This is called the peptide bond, -CO-NH-. Polymerization of amino acids, by the formation of a series of such bonds, probably took place, producing elongate chains, like protein. These protein-like compounds have been termed protenoids. In brines in the laboratory, protenoids tend to concentrate in microspheres, which are globules about 1 $\mu$m (one thousandth of a millimetre) in diameter. Although abiotic, microspheres do have some of the properties of division and metabolism of living cells. However, there is still a gap between microspheres and simple cells, because the former neither replicate nor produce energy by feeding or photosynthesis as do true living cells.

During the early history of the Earth, then, compounds with complex, interwoven chains of carbon atoms presumably became synthesized. Later, such compounds may have been broken by the severing of chemical bonds along the length of each chain. And when the broken strands acted as templets on which new strands grew, to be broken off anew, life had essentially arrived. The complexity of the process is staggering, for, to make the first living cell, a host of compounds must have been available, yet, individually, each must have been useless. What good is cell membrane material that does not enclose the contents of a cell? But, perhaps surprisingly, life evolved.

# Glossary

**Ablation:** Removal of material, for example, from an object entering the atmosphere from space. In this case friction with the atmosphere causes the surface of the object to melt or to boil and hence to be stripped off.

**Accretion:** Growth by addition of material, perhaps by the collision and sticking together of particles into an aggregate, or by the accumulation of material on a planetary surface, thereby causing the planetary body to grow in size.

**Alpha-particle ($\alpha$):** The nucleus of a helium 4 atom, comprising two protons and two neutrons and of atomic weight 4. An extremely stable nuclear structure which takes part in nuclear reactions.

**Annealing:** Production of change in crystal structure during slow cooling.

**Antimatter:** The opposite of matter as we know it, for example a 'proton' with a negative electric charge (antiproton). Collision of a proton (which has a positive electric charge) with an antiproton results in annihilation, with release of energy.

**Asteroid:** A small planetary body that orbits the Sun, generally between Mars and Jupiter. The largest asteroid, Ceres, has a diameter of 1025 km; that of the Moon is 3500 km.

**Asteroid belt:** The zone between Mars and Jupiter in which the bulk of the asteroids are located. Belt asteroids have fairly circular orbits which do not approach the orbits of Mars and Jupiter too closely.

**Atoms:** The smallest particles of matter, with no electrical charge, which take part in chemical reactions. Each atom has a tiny nucleus, with a positive electric charge, and one or more electrons in orbit around it. The number of the negative electric charges of the electrons balances the number of protons in the nucleus. The charge of the proton is exactly the opposite of the charge of the electron. The proton is much heavier than the electron. In addition to protons, the atoms of all elements except hydrogen have one or more neutrons in the nucleus. The neutron weighs the same as the proton, but has no electric charge.

**Atomic symbols:** A list of symbols for most of the chemical elements mentioned in the text is shown on the following page.

**Beta-particle ($\beta$):** A free electron.

**Black hole:** Dense, concentrated matter with gravity so great that light cannot escape.

**Catalyst:** A substance whose presence aids a chemical reaction but which is not consumed by the reaction.

## Light elements

| | | | | |
|---|---|---|---|---|
| H | Hydrogen | | He | Helium |
| C | Carbon | | N | Nitrogen |
| O | Oxygen | | F | Fluorine |
| Ne | Neon | | Na | Sodium |
| Mg | Magnesium | | Al | Aluminium |
| Si | Silicon | | P | Phosphorus |
| S | Sulphur | | Cl | Chlorine |
| Ar | Argon | | K | Potassium |
| Ca | Calcium | | Ti | Titanium |
| Cr | Chromium | | Fe | Iron |
| Ni | Nickel | | Ga | Gallium |
| Ge | Germanium | | Br | Bromine |
| Rb | Rubidium | | Sr | Strontium |
| I | Iodine | | Xe | Xenon |
| Pt | Platinum | | Au | Gold |
| Pb | Lead | | U | Uranium |
| Pu | Plutonium | | | |

## Heavy elements

**Comet:** A body of low mass, composed largely of ices. Comets possibly populate the outermost regions of the solar system. They orbit the Sun but are invisible to man unless their orbits are changed by the action of gravity (a passing star?) sending them into the inner solar system. On a close approach and passage around the Sun, heating causes some of the ices to vaporize, which releases dust and gas to form a tail, or tails. It is the effect of solar radiation that renders a comet visible.

**Cosmic rays/radiation:** Particles of matter travelling at a speed that is a significant fraction of the speed of light, hence their great energy. Their origin is unknown, but probably within our galaxy.

**Cosmic radiation age:** Time during which a solid object, or surface, is bombarded by cosmic rays. Cosmic rays interact with matter, so our atmosphere acts as a screen against most cosmic radiation. Most cosmic rays produce nuclear reactions to a depth of some tens of centimetres in stony material, so the cosmic radiation age is the time between the ejection of a metre-sized object from a parent-body, and its fall to Earth as a meteorite, when it is again screened.

**Escape velocity:** The velocity to which an object has to be accelerated to leave a planet or satellite. Conversely, it is the velocity to which a 'stationary' object at infinity would be accelerated on hitting the surface of a body with no atmosphere. The Earth's escape velocity is 11.2 km per second, that of Mars, 5 km per second.

**Fall:** A meteorite recovered soon after its arrival on Earth, mainly due to observation while it was in atmospheric flight.

**Find:** A meteorite discovered an indeterminate time since it landed on Earth.

**Fireball:** Moving bright light in the atmosphere caused by frictional heating of an incoming object.

**Fission:** Spontaneous splitting of an atom into two (or more) fragments of approximately equal weight, which fly apart with great energy.

**Fractionation:** Production of a mass of material that is (chemically) distinct from a larger mass of parental material. For example, to produce pure water by distillation of brine entails vapour fractionation.

**Fusion:** Combination of atomic nuclei lighter than the iron nucleus, with the evolution of energy.

**Isochron:** A line on a diagram used in absolute age determination. If data from a number of rocks or meteorites (or from their separated minerals) plot on an isochron, then the rocks or meteorites have the same age, which is proportional to the slope of the line. (Technically, for each sample, the ratio of the abundance of a radiogenic isotope to that of a non-radiogenic isotope of the same element is plotted against the ratio of the abundance of the radioactive parent isotope to that of the same non-radiogenic isotope, e.g. $^{87}Sr/^{86}Sr$ is plotted against $^{87}Rb/^{86}Sr$.

**Isotope:** An atom of an element with a specified number of neutrons in its nucleus. Most hydrogen atoms have no neutrons. However, heavy hydrogen, or deuterium, has a neutron in its nucleus in addition to the single proton. Thus hydrogen has two stable isotopes, one of which (deuterium) is twice as heavy as the other, although they are chemically identical.

**Kinetic energy:** The energy which an object has by virtue of its motion. Kinetic energy is the product of the mass of the object and the square of its velocity. Thus, a 100 tonne object striking the Earth at 1 km per second has the same energy as a 1 tonne object travelling at 10 km per second. ($100 \times 1 \times 1 = 1 \times 10 \times 10$.)

**Light year:** An astronomical unit of distance; the distance that light travels in a vacuum in one year, 9 460 000 000 000 km, or 63 240 times the mean distance of the Earth from the Sun.

**Mare:** Large lunar basin filled or partially filled with dark lavas.

**Mascon:** A concentration of dense matter beneath some maria (the plural of 'mare'), which produces local enhancement of lunar gravity.

**Mass-spectrometer:** A device for measuring the abundances of isotopes of a chemical element. Because the isotopes of an element differ in their weights but are chemically identical, they can be distinguished only by physical means. The atoms are heated and become electrically charged. They are then accelerated, under vacuum, and passed through a magnetic field, which causes the paths of the charged atoms to bend. Light isotopes are deflected more than heavy ones for a given acceleration and magnetic field. By changing the magnetic field and/or acceleration, the scientist can detect charged atoms of a specific isotope in a suitably positioned collector.

**Meteor:** Transient phenomena caused by the entry of an object into the upper atmosphere. At night, a meteor may be seen as a shooting star, a bright moving light which may leave a luminous trail.

**Meteorite:** A natural object which survives its fall to Earth from space.

**Oxidize:** The process by which the atoms of an element lose electrons. When iron rusts, the metal loses electrons to oxygen and water, which gain electrons, as the iron combines with them and becomes oxidized. The oxygen and water, in gaining electrons, are said to be reduced. The processes of oxidation and reduction go hand in hand.

**Polymer:** A chemical compound with a chain structure caused by the linking of many identical units.

**Polymerization:** The process by which simple molecules (units) become linked. For example, the gas ethylene has two carbon atoms each in combination with two hydrogen atoms: $H_2C = CH_2$. Loss of hydrogen can cause the carbon atoms to combine with other units:

$$\underset{\displaystyle \overset{\textstyle |}{H}}{-C} = \overset{\displaystyle \overset{\textstyle H}{|}}{C} - \overset{\displaystyle \overset{\textstyle |}{H}}{C} = \overset{\displaystyle \overset{\textstyle H}{|}}{C} - \overset{\displaystyle \overset{\textstyle |}{H}}{C} = \overset{\displaystyle \overset{\textstyle H}{|}}{C} - \quad \text{forming polyethylene (polythene).}$$

**Quench:** To cool rapidly from high temperature, which inhibits or modifies crystallization.

**Radioactive:** An element is said to be radioactive if a proportion of its atoms transmute naturally in a given time to atoms of another element, or elements, with the release of energy.

**Radiogenic:** Describes the (daughter) atoms produced by a radioactive parent.

**Reduce:** see 'Oxidize'.

**Refractory:** Resistant to high temperature; a substance which is difficult to melt.

**Shooting star:** see under 'meteor'.

**Solar wind:** Particulate radiation from the Sun. Atomic nuclei, accelerated by an unknown mechanism, stream out from the Sun. The particles are of low energy and are dominantly protons (see also under 'Atoms').

**Supernova:** A star which explodes, observed as a burst of light which gradually diminishes. This is followed by the ejection of matter processed in the explosion.

**Tektite:** Natural glassy objects confined to particular areas of the Earth's surface and apparently not associated with volcanism. In some cases (e.g. FIG 6.5, p. 120; specimens from Victoria, Australia) their form indicates that they were ablated during hypersonic flight. Trace element and isotopic abundances testify to a terrestrial origin. There is now a consensus among scientists (with few exceptions) that tektites are parts of the Earth's crust melted or vaporized by a major impact. Condensation and/or solidification occurred outside the atmosphere, and ablation occurred during re-entry.

**Vaporize:** To transform from solid or liquid to a gas.

**X-rays:** Electro-magnetic radiation of wavelength shorter than ultraviolet (U.V.) at the limit of visual range. X-rays are more energetic than U.V. or visual light.

# Index and Reading lists

## Background reading

Betty, J. K., O'Leary, B. and Chaikin, A. (editors). 1981. *The new solar system.* 232 pp. Cambridge University Press, Cambridge.

Calder, N. 1980. *The comet is coming.* 160 pp. BBC, London.

Clark, D. H. 1979. *Superstars.* 175 pp. Dent, London.

Cocks, L. R. M. (editor). 1981. *The evolving Earth.* 264 pp. British Museum (Natural History) and Cambridge University Press, London and Cambridge.

French, B. M. 1977. *The Moon book.* 287 pp. Penguin, Harmondsworth.

## Further reading

Dodd, R. T. 1982. *Meteorites: a petrologic-chemical synthesis.* 368 pp. Cambridge University Press, Cambridge.

King, E. A. 1976. *Space geology.* 349 pp. Wiley, New York.

Short, N. M. 1975. *Planetary geology.* 361 pp. Prentice Hall, Englewood Cliffs, New Jersey.

Smith, J. V. 1979. Mineralogy of the planets: a voyage in space and time. *Mineralogical Magazine,* 43; 1-89.

Taylor, S. R. 1982. *Planetary science: A lunar perspective.* 512 pp. Lunar and Planetary Institute, Houston.